Maurice Norton Miller, Herbert Upham Williams

Student's Histology

Maurice Norton Miller, Herbert Upham Williams

Student's Histology

ISBN/EAN: 9783337399191

Printed in Europe, USA, Canada, Australia, Japan

Cover: Foto ©berggeist007 / pixelio.de

More available books at **www.hansebooks.com**

STUDENTS HISTOLOGY

*A COURSE OF NORMAL HISTOLOGY
FOR STUDENTS AND
PRACTITIONERS OF MEDICINE*

BY

MAURICE N. MILLER, M.D.

Late Director of the Department of Normal Histology in Loomis' Laboratory,
University of the City of New York

REVISED BY

HERBERT U. WILLIAMS, M.D.

Professor of Pathology and Bacteriology, Medical Department,
University of Buffalo

**THIRD REVISED EDITION
PROFUSELY ILLUSTRATED**

NEW YORK
WILLIAM WOOD & COMPANY
1898

PREFACE

This volume has been prepared with a view of aiding the instructors and students of the laboratory classes which are under my direction.

It is also presented with the hope that it may be useful to other instructors.

Again, students often wish to continue microscopical work during the interim of college attendance; to such, it is my belief, these pages will have some value.

Still again, very many practitioners, not having had, during pupilage, advantages equal to those provided by the modern laboratory equipment, wish to acquire more knowledge of microscopy, for its value in practical medicine. To such workers, also, I desire to be useful.

So much technique has been introduced as has been found to be of absolute necessity, and no more. The processes for the preparation and exhibition of tissues are generally simple and always practicable.

In the description of organs, I assume that the student has a fair knowledge of gross anatomy, but knows nothing of histology. The scheme or plan of the structure is first described —using diagrams where requisite to clearness—after which the mode of preparing the sections is indicated, and, under practical demonstration, every histological detail tabulated in proper order. The drawings will, I believe, aid in the recognition of such elements in the field of the microscope.

The illustrations are exact reproductions, by photography, of my own pen-pictures; and distinction must always be made between the drawings which are schematic—used to emphasize the plan of structures — and those drawn from the tissue as seen in the microscope.

Our literature abounds in excellent works for the advanced student, and this volume is designed to pave the way for their appreciation.

I desire to record my high appreciation of the aid of Drs. Charles T. Jewett, Egbert Le Fevre, E. Eliot Harris, Milton Turnure, H. Pereira Mendes, J. Gorman, A. M. Lesser, J. Alexander Moore, Robert Roberts, Esq., Warden, and Mr. John Burns, Clerk of Charity Hospital, in facilitating my access to valuable tissue for the illustrations and for my own studies.

My thanks are due my First Assistant, Dr. F. T. Reyling, for his indefatigable efforts in furthering the work; and to Mr. A. J. Drummond, for photographical favors.

MAURICE N. MILLER.

NEW YORK, June 1st, 1887.

PREFACE TO THIRD REVISED EDITION

A revision of Miller's Microscopy became necessary, partly on account of the advances made in histology during the last ten years, and partly because of the increasing tendency in medical schools to devote more time to laboratory studies. Substantially all of the original matter has been retained, although somewhat rearranged; and where new matter has been inserted, the attempt has been made to give it the form which was the peculiarity of the original, namely, in being written from the point of view of the student, and not of the teacher. In preparing the various additions, the standard text-books on histology have been constantly consulted.

BUFFALO, N. Y., August, 1898.

CONTENTS

PART FIRST

TECHNOLOGY

THE LABORATORY MICROSCOPE

	PAGE
Description of the Stand	1
Lenses	2
General Adjustment	4
Adjustment for Illumination	4
Adjustment for Focus	5
Method in Observation	6
Conservation of the Eyesight	7
Magnifying Power	7
Measurement of Objects	8
Sketching from the Instrument	9

PREPARATION OF TISSUES FOR MICROSCOPICAL PURPOSES

Teasing of Tissues	9
Section Cutting	10
Free-hand Section Cutting	10
Section Cutting with the Stirling Microtome	12
Section Cutting with the Larger Microtomes	15
Sharpening Knives—Honing and Stropping	16
Supporting Tissues for Cutting	18
Paraffin Soldering	18
The Freezing Microtome	19
Fixing or Hardening Fluids	20
Alcohol Hardening	21
Müller's Fluid	22
Orth's Fluid	22
Formaldehyde	22
Picric Alcohol	23
Osmic Acid	23
Flemming's Solution	23
Chromic Acid	24
Erlicki's Fluid	24

(v)

CONTENTS

	PAGE
Decalcifying Fluid	24
Dissociating Fluid	25
Imbedding—Paraffin	25
" —Celloidin	26
Staining Methods in General	27
Hæmatoxylin	28
Eosin	28
Carmine	29
Weigert-Pal Method	30
Aniline Dyes	31
Van Gieson's Stain	32
Ehrlich Tricolor Stain	32
Metallic Impregnations—Nitrate of Silver	33
Golgi's Method	33
Injection Methods	34
Clearing Agents	35
Mounting Media	35
Hæmatoxylin Staining Process	36
Hæmatoxylin and Eosin Double Staining	38
Borax-Carmine Staining Process	39
Cleaning Slides and Cover-glasses	41
Mounting Methods	41
Labeling Slides	42
Care of the Microscope	43

PART SECOND

STRUCTURAL ELEMENTS

PRELIMINARY STUDY

Form of Objects	45
Movement of Objects	46
Extraneous Substances	47

STRUCTURAL ELEMENTS

Cells	49
Cell Distribution	50
Cell Division—Karyokinesis	51
Classification of Tissues	53
Embryonic Derivation of Tissues	54
Epithelium	54
Distribution of Epithelium	55
Squamous, Stratified, and Transitional Epithelium	55
Pavement Epithelium	56
Columnar Epithelium	57
Ciliated Columnar Epithelium	58

CONTENTS vii

	PAGE
Glandular Epithelium	59
Endothelium—Serous Membranes	60

CONNECTIVE (FIBROUS) TISSUES

White Fibrous Tissue	62
Yellow Elastic Tissue	63
Adipose Tissue	65

CARTILAGE

Hyaline Cartilage	67
White Fibro-Cartilage	68
Elastic Cartilage	68

BONE

Bone	70
Periosteum	73
Marrow	73
Development of Bone	73

SPECIAL CONNECTIVE TISSUES (page 76)

MUSCULAR TISSUE

Non-striated Muscle	76
Striated Muscle	78
Cardiac Muscle	80

BLOOD

Red Corpuscles	81
Blood-plates	82
White Corpuscles	83
Enumeration of Blood-corpuscles	87
Hœmoglobin	89
Fibrin	90
Effect of Reagents	90
Development of Red Blood-corpuscles	91

PART THIRD

ORGANS

THE SKIN

Layers or Strata	92
Hairs	95
Sudoriferous Glands	97

	PAGE
Sebaceous Glands	98
Nails	98
Practical Demonstration	98

THE CIRCULATORY SYSTEM

The Heart	102
Blood-vessels	102
Development of Capillaries	105

THE LYMPHATIC SYSTEM

General Description	106
Lymph-channels	107
Practical Demonstration, Lymph-channels of Central Tendon of the Diaphragm	108
Lymphatic Nodes or Glands	111
Practical Demonstration, Mesenteric Lymph-node	113

THE SPLEEN

Scheme of Organ	117
Practical Demonstration	118

THE THYMUS BODY

General Description	121
Practical Demonstration	121

THE RESPIRATORY ORGANS

The Larynx and Trachea	123
The Lungs	123
Bronchial Tubes	123
Practical Demonstration	125
Pulmonary Blood-vessels	128
Pleura	128
Pulmonary Alveoli	128
Practical Demonstration, Lung of Pig	131
Practical Demonstration, Human Lung	133
Fœtal Lung	134

THE TEETH

The Pulp	135
Dentine	135
Enamel	137
Cementum	137
Practical Demonstration	137

GLANDS

	PAGE
Typical Glandular Histology	140
Tubular Glands	140
Coiled Tubular Glands	140
Branched Tubular Glands	143
Acinous Glands	143
Parotid Gland	143
Submaxillary Gland	144
Pancreas	145
Practical Demonstration, Parotid Gland, Submaxillary Gland, Pancreas	146
Thyroid Gland	148

THE ALIMENTARY CANAL

The Mouth and Pharynx	149
The Œsophagus	150
Practical Demonstrations	150
General Histology of the Stomach and Intestines	151
The Stomach	152
Practical Demonstration	155
Small Intestine	156
Practical Demonstration	160
Duodenum, Vermiform Appendix, Large Intestine	162

THE LIVER

General Scheme	163
The Portal Canals	165
The Lobular Parenchyma	165
Practical Demonstration, Liver of Pig	167
Practical Demonstration, Human Liver	170
Practical Demonstration, The Portal Canals	172
Practical Demonstration, The Lobular Parenchyma	173
Practical Demonstration, Origin of Bile-ducts	176
Gall-bladder	177

THE KIDNEY

General Description	178
The Tubuli Uriniferi	180
Blood-vessels	183
Practical Demonstration, with Low-power	186
Practical Demonstration, The Cortical Portion	187
Practical Demonstration, The Medullary Portion	192
Pelvis of the Kidney and Ureter	193
Urinary Bladder	194
Urethra	196

FEMALE GENERATIVE ORGANS

	PAGE
Vagina and Uterus—Practical Demonstration	197
Fallopian Tube—Practical Demonstration	202
Ovary	203
Adult Human Ovary—Practical Demonstration	203
Formation of the Ovum	206
Ovary of Human Infant—Practical Demonstration	206
Mammary Gland	208

MALE GENERATIVE ORGANS

Testicle—General Description	209
Practical Demonstration	211
Spermatozoa	211
Prostate Gland	212
Erectile Tissue	212

SUPRARENAL BODY

Practical Demonstration	213

THE NERVOUS SYSTEM

Structural Elements	216
Nerve-fibers	216
Nerve-trunks	217
Practical Demonstration	219
Nerve-cells	219
Neuroglia	222
Peripheral Nerve-endings	223

THE SPINAL CORD

General Description	226
Division into Tracts	228
Practical Demonstration	229
Relation of Ganglion-cells and Nerve-fibers	233

THE BRAIN

Membranes	235
Cerebrum—Practical Demonstration	237
Layers of the Cerebral Cortex	238
Paths followed by Nerve-fibers in the White Matter	239
Cerebellum—Practical Demonstration	240

INDEX (page 245)

LIST OF ILLUSTRATIONS

FIG.		PAGE
1.	Microscope	2
2.	Relation of Objective to Eye-piece	3
3.	English and Metric Scales	8
4.	Free-hand Section Cutting	11
5.	Stirling's Microtome	12
6.	Method of Imbedding with Pith, Turnip, etc.	13
7.	Section Cutting with Stirling Microtome	14
8.	Thoma Microtome	14
9.	Schanze Microtome	15
10.	Method of Honing Razor	16
11.	Turning the Razor on the Hone	16
12.	Paraffin Soldering Wire	18
13.	Cementing Hardened Tissue to Cork	18
14.	Freezing Microtome	19
15.	Using Turn-table	36
16.	Needle for Lifting Sections	36
17.	Diagram Illustrating Steps in Staining with Hæmatoxylin	37
18.	Diagram Illustrating Steps in Staining with Hæmatoxylin and Eosin	39
19.	Diagram Illustrating Steps in Staining with Borax-Carmine	40
20.	Section-lifters	41
21.	Appearance of Balsam-mounted Specimen	42
22.	Mode of Handling Cover-glass	42
23.	Diagram Showing Effect of Oil and Air Globules	46
24.	Extraneous Substances—Fibers, etc.	47
25.	Extraneous Substances—Starch, etc.	48
26.	Elements of a Typical Cell	49
27.	Structure of a Cell-nucleus	50
28.	Indirect Cell Division	51
29.	Karyokinesis	52
30.	Squamous Epithelial Cells from Mouth	56
31.	Pavement Epithelium	57
32.	Columnar Cells from Intestine	58
33.	Ciliated Columnar Cells from Bronchus	58
34.	Diagram Showing Organs of the Oyster	59
35.	Glandular Cells from Liver	60
36.	Frog's Mesentery—Silver-staining	61
37.	White Fibrous Tissue	63
38.	Yellow Elastic Tissue	64
39.	Transverse Section of Ligamentum Nuchæ	64

xii LIST OF ILLUSTRATIONS

FIG.		PAGE
40.	Cells containing Fat	66
41.	Adipose Tissue from Omentum	66
42.	Hyaline Cartilage from Bronchus	67
43.	Fibro-Cartilage from Intervertebral Disc	68
44.	Elastic Cartilage from Ear of Bullock	69
45.	Bone—Showing Laminated Structure	69
46.	Bone—Showing Haversian Systems	70
47.	Bone—Showing Sharpey's Fibers	71
48.	Contents of Haversian Canals	71
49.	Contents of Bone Lacuna	72
50.	Cells from Red Marrow	74
51.	Developing Bone	75
52.	Non-striated Muscular Fiber	77
53.	Striated Muscular Fiber	78
54.	Striated Muscular Fiber from Tongue	79
55.	Cardiac Muscular Fiber	80
56.	Corpuscular Elements of Human Blood	81
57.	Diagram of Red Blood-corpuscle	82
58.	Blood Showing Blood-plates	83
59.	Cover-glass, Preparations of Blood	84
60.	Varieties of Leucocytes	85
61.	Pipettes for Hæmocytometer	86
62.	Disk of Hæmocytometer	87
63.	Magnified Field of Hæmocytometer	88
64.	Crystals of Hæmoglobin	89
65.	Blood-corpuscles of the Frog	90
66.	Layers of the Epidermis	93
67.	Structure of the Derma—Injected	94
68.	Hair in Transverse Section	95
69.	Hair Follicle	96
70.	Sudoriferous Gland	97
71.	Sebaceous Gland	98
72.	Skin in Vertical Section	100
73.	Diagrammatic Section of Artery	103
74.	Blood-capillaries	104
75.	Perivascular Lymph-spaces	107
76.	Lymphatics of Central Tendon of the Diaphragm—Low-power	110
77.	Lymphatics of Central Tendon of the Diaphragm—High-power	111
78.	Lymph-node—Diagrammatic	112
79.	Mesenteric Lymph-node—Low-power	114
80.	Mesenteric Lymph-node—High-power	115
81.	Blood-vessel Arrangement in the Spleen	117
82.	Spleen	119
83.	Thymus Body	122
84.	Bronchial Tube—Small	124
85.	Bronchial Tube—Medium	127
86.	Pulmonary Lobule—Perspective	129
87.	Pulmonary Lobule—Longitudinal Section	129
88.	Pulmonary Alveolus—Capillaries Injected	130

LIST OF ILLUSTRATIONS

FIG.		PAGE
89.	Lung of Pig	132
90.	Pulmonary Alveolus Showing Lining	133
91.	Diagrammatic Section of Tooth	136
92.	Section of Part of Tooth—High-power	139
93.	Simple Tubular Gland	141
94.	Coiled Tubular Gland	141
95.	Branched Tubular Gland	142
96.	Acinous Gland	142
97.	Parotid Gland	144
98.	Submaxillary Gland	145
99.	Pancreas	146
100.	Stomach—Diagrammatic Section	152
101.	Cardiac Gastric Gland	153
102.	Pyloric Gastric Gland	154
103.	Stomach of Dog	155
104.	Diagram Illustrating Intestinal Secretion	158
105.	Diagram of Intestinal Absorption	159
106.	Small Intestine with Peyer's Patch	161
107.	Liver—Diagram Illustrating Plan of Structure	164
108.	Glandular Cells in Connection with Blood-vessels and Ducts	166
109.	Liver of Pig	168
110.	Human Liver—Low-power	171
111.	Portal Canal	173
112.	Hepatic Cells—Detached	174
113.	Hepatic Lobule in Transverse Section	175
114.	Bile-capillaries—Origin of Bile-duct	176
115.	Kidney—Diagram Illustrating Plan of Structure	179
116.	Kidney Tubules—Isolated	181
117.	Blood-vessels—Arrangement in Kidney	183
118.	Kidney—Low-power	186
119.	Kidney—Cortex in Vertical Section	188
120.	Kidney—Medulla in Longitudinal Section	191
121.	Kidney—Medulla in Transverse Section	192
122.	Epithelium of Ureter	194
123.	Epithelium of Urinary Bladder	195
124.	Uterus with Vaginal Cul-de-sac	198
125.	External Os Uteri	200
126.	Vaginal Epithelium	201
127.	Fallopian Tube	202
128.	Ovary—Adult	204
129.	Ovary—Child's	207
130.	Mammary Gland—Dog	208
131.	Mammary Gland—Dog	208
132.	Testicle, Diagram	209
133.	Seminiferous Tubules	210
134.	Spermatozoa	211
135.	Suprarenal Body—Low-power	214
136.	Suprarenal Body—High-power	215
137.	Nerve-fibers	216

LIST OF ILLUSTRATIONS

FIG.		PAGE
138.	Nerve-trunk—Transverse Section	218
139.	Two Types of Ganglion-cells	220
140.	Diagram of a Neurone	221
141.	Neuroglia	222
142.	Nerve-endings in the Cornea	223
143.	Tactile Corpuscle	224
144.	Pacinian Corpuscle	224
145.	Nerve-ending in Striated Muscle	225
146.	Diagram, Spinal Cord	226
147.	Dorsal Spinal Cord	227
148.	Lumbar Spinal Cord	228
149.	Cervical Spinal Cord	230
150.	Anterior Horn—Gray Matter—Cervical Spinal Cord	231
151.	Ganglion cells, Anterior Horn	233
152.	Diagram, Relations of Cells and Nerve-fibers in the Spinal Cord	234
153.	Layers of the Cerebral Cortex	236
154.	Section of Cerebrum	237
155.	Ganglion-cell and Neuroglia-cell—Cerebral Cortex	239
156.	Cerebellum—Low-power	240
157.	Cerebellum—High-power	241
158.	Cell of Purkinje	242

STUDENTS HISTOLOGY

Part First

TECHNOLOGY

THE LABORATORY MICROSCOPE

The histologist should be provided with a microscope, in which the principal features of the laboratory instrument, Fig. 1, are embraced.

The body A, which carries the optical parts, is made of two pieces of brass tubing, one sliding within the other and providing for alterations in length. The *objectives*, B, C, D, are attached to the body by means of the triple nose-piece, E. The nose-piece is so pivoted that either objective may be turned into the optical axis at will. The *eye-piece*, F, slips into the upper part of the body with but little friction, so that it may be quickly and easily removed.

The *coarse* or *quick adjustment* for focusing consists of a rack, G, which is attached to the body, and into this gears a small (concealed) pinion turned by the milled heads, H.

The more delicate adjustments are accomplished by means of a micrometer screw acting by a simple mechanical device upon an enclosed spring in connection with a prism slide. By turning the milled head, L, the body of the instrument is raised or lowered, as desired, and with extreme delicacy.

The *stage*, M, upon which objects are placed for examination, is perforated, and an iris diaphragm and Abbé condenser, K, may be inserted. The iris diaphragm enables one to alter the size of the opening at will. Below the stage an arm may be seen which carries a sliding fork supporting the mirror, N, one side of which is plane and the other concave.

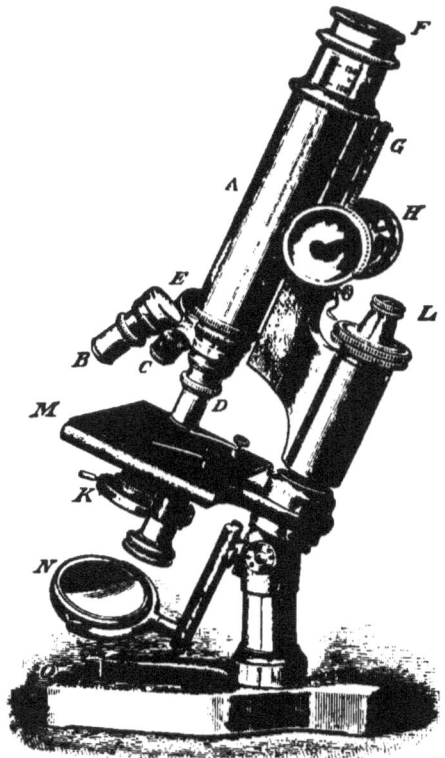

Fig. 1. The Microscope.

The whole is supported on a short, stout pillar rising from the base, O.

LENSES OF THE MICROSCOPE

Fig. 2 shows the arrangement of lenses, including a high-power objective.

The *objective*, A, is provided with one simple and two compound lenses. The lens, B, nearest the object, and the one upon which the magnifying power mainly depends, is a hemisphere of crown glass. A lens of such form possesses both chromatic and spherical aberration in high degree. These faults are corrected by the compound flint and crown lenses, C and D, placed above the hemispherical glass.

The *eye-piece* consists of two crown glass, plano-convex lenses, with their plane surfaces upward. The lower, E, is known as the field-lens; the upper, F, as the eye-lens. Eye-pieces add very materially to the magnifying power of the instrument, and are

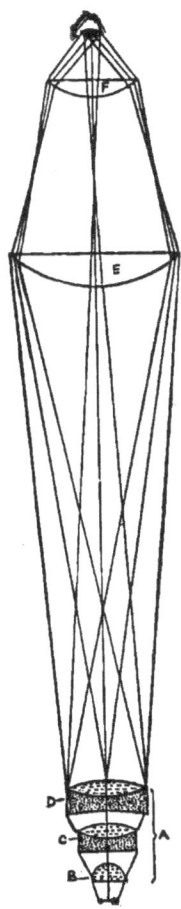

Fig. 2. Diagram Showing the Relation of the Objective to the Eye-piece.

constructed of various strengths depending upon the curvature of the lenses. They are named according to power, A, B, C, or 1, 2, and 3, or according to their focal distances. The medium power is more commonly employed.

The microscope previously described stands, with the draw-tube in place, about twelve inches high. If the height of the table upon which it is placed and the chair of the observer be in a proper relation, no discomfort need be experienced in using the microscope in the vertical position. The form of condenser invented by Abbé may be placed directly below the stage (Fig. 1, K). It consists of two or three lenses combined so as to focus the rays coming from the *plane* mirror upon the object. The condenser gives a very intense illumination over a small field, and is adapted to bacteriological work, where a high-power oil-immersion objective is required. An oil-immersion objective is a specially constructed system of lenses, with which a layer of thickened oil of cedar wood is placed between the lower surface of the objective and the upper surface of the glass covering the object under examination. The oil-immersion lens in general use has an equivalent focal length of one-twelfth of an inch, and is usually designated as the $\frac{1}{12}$-inch oil-immersion. For similar reasons, the low-power is a $\frac{2}{3}$- or $\frac{3}{4}$-inch, and the ordinary high-power is a $\frac{1}{4}$-, $\frac{1}{5}$-, or $\frac{1}{6}$-inch objective, as the case may be. The condenser is not necessary, except with the high-power oil-immersion objective. If used with the other objectives, the illumination must be regulated by lowering the condenser, closing the diaphragm, and substituting the concave for the plane mirror, till a clear and satisfactory picture is secured.

ADJUSTMENT OF THE MICROSCOPE

The microscope should be placed in front of the observer, on a table of such height that, when seated, he may, by slightly inclining the head, and without bending the body, bring the eye easily over the eye-piece. The slightest straining of the body or neck should be avoided. The light should always be taken from the side, and it matters little which side. Clouds or clear sky serve as the best source of light for our present work. Always avoid direct sunlight. If artificial illumination be employed—though it is not advised for prolonged investigation—a small coal-oil flame may be tempered by blue glass, or better a Welsbach gas-burner with a blue glass globe, or an incandescent electric light.

ADJUSTMENT FOR ILLUMINATION

It will be observed that there are two mirrors in the circular frame below the stage—one plane and the other concave. The latter will be employed almost exclusively in the work of this volume, and its curvature is such that parallel rays, impinging upon its surface, are focused about two inches from the mirror. It will also be noticed that the bar, carrying the mirror-fork, may be made to swing the mirror from side to side. The work which we are about to undertake is of such a character as to require the avoidance of oblique illumination. We must, therefore, keep our mirror-bar strictly in the vertical position. If—the mirror-bar being vertical—a line be drawn from the center of the face of the mirror, through the opening (diaphragm) in the stage, passing on through the objective, and so continued upward through the body and the eye-piece, such a line would pass through the *optical axis*. The center of the face of the mirror must be in this axis. If, then, having gotten the mirror-bar properly fixed once for all, the light from the adjacent right or left hand window impinges upon the concave surface of the mirror, and the latter be properly inclined, the rays will pass through the diaphragm in the stage, and become focused a little above the same. The light rays will afterward diverge, enter the objective, and finally reach the eye of the observer.

The *field* of view (as the area seen in the microscope is termed) we will suppose to have been properly illuminated—and by this we mean that it presents us a clear, evenly lighted area. Turn all the factors spoken of out of adjustment, and proceed to readjust. Observe that, if the mirror be turned—not swung—slightly out of proper position, one side of the field will appear dim or cloudy. This must be corrected, and the student must practice until this adjustment becomes easy of accomplishment. Then proceed to the

ADJUSTMENT FOR FOCUS

Swing the low-power objective into use, and rack the tube up or down until it is about one-fourth of an inch from the stage. Place a mounted object upon the stage (a stained section of some organ—say kidney—will be preferable). Examine the field through the eye-piece, and it will be found obscured by the stained object, and perhaps a dim notion of figure may be made out. Rack the body up carefully, watching the effect. The image becomes more and more distinct until, at a certain point, the best effect is secured. The object is *in focus*.

Note carefully the distance between the object and the objective (with the low-power this will be less than one-half of an inch), and hereafter you will be able to focus more quickly.

Having observed the details of structure as shown with the low-power, swing the high-power into use. Rack the tube down until the objective is very near to the glass covering the object. The field is much obscured. Watching the effect through the eye-piece, rack the tube up with great care until the image appears sharp. Note the distance with this objective, as before with the low-power, from one-twelfth of an inch to considerably less. Then endeavor, by slight alterations in the inclination of the mirror, to increase the illumination. Turn the diaphragm so that the light passes through a small opening, and note the improvement in definition. The rule is: *The higher the power, the smaller the diaphragm.*

You have doubtless observed, before this, that you cannot control the focusing as easily as when the low-power was in use. Slight movements of the rack-work produce marked changes in definition; and it is difficult, with the coarse adjustment alone, to make as slight movements as you may desire. Recourse must be had to the fine adjustment.

Place the tip of the forefinger (either) upon the milled head of the fine focusing-screw, and the ball of the thumb against its side, so that the hand is in an easy position. By a little lateral pressure the milled head may be turned slightly either way. Note the effect on the image. You thus have the focusing under the most perfect control.

Remember that the fine adjustment is only necessary with high-powers, and then only after the image has been found with the coarse adjustment.

METHOD IN OBSERVATION

The study of objects under the microscope should be conducted with order and method.

The body being in the position before advised, so that the sitting may be prolonged without fatigue, let one hand be occupied in the maintenance of the focal adjustment. It will be found, however flat an object may seem to the unaided eye, that as it is moved so as to present different areas for examination (and with the higher-powers only a small area can be seen at once), constant manipulation with the fine adjustment will be required. It will also be found that even the various parts of a simple histological element—like a cell—cannot be seen sharply with a single focal adjustment. The forefinger and thumb of one hand must be kept constantly on the milled head of the fine focusing-screw. Supposing the light to be on our right, we devote the right hand to the focusing.

The left hand will be engaged with the glass slip upon which the object has been mounted. The forearm resting upon the table, let the thumb and forefinger rest on the left upper side of the stage, just touching the edges of the glass slip. The slightest pressure will then enable you to move the slip smoothly, steadily, and delicately.

Proceed to examine the object with method. Suppose a section of some tissue to be under examination—say one-fourth of an inch square. With the high-power you will be able to see only a small fraction of the area at once. Commence at one corner to observe, and, with the left hand, move the object slowly in successive parallel lines, preserving the focus with the right hand, until the whole area of the section has been traversed.

Practice will soon establish perfect co-ordination of the movements involved, and will result in the ability to work with ease, celerity, and profit.

CONSERVATION OF THE EYESIGHT

The beginner should not become accustomed to the use of one eye alone, or of closing either in microscopical work. It will require but little practice to use the eyes alternately, and the retinal image of the unemployed eye will soon be ignored and unnoticed.

MAGNIFYING POWER AND MEASUREMENT OF OBJECTS

The microscope is not, as the beginner usually supposes, to be valued according to its power of enlargement or magnification, but rather according to the clearness and sharpness of the image afforded.

Magnifying power is generally expressed in diameters. A certain area is by the instrument made to appear, say, ten times as large as it appears to the naked eye. This object, then, has its apparent area increased one hundred times; but reference is made in describing such phenomena only to amplification in a single direction. The diameter of the object under examination has been increased ten times and would be expressed by prefixing the sign of multiplication: *e.g.*, × 10.

A convenient unit of approximate measurement for the histologist is the apparent size of a human red blood-corpuscle with a given objective. Thirty-two hundred corpuscles, placed side by side, would measure one inch; or, we say, the diameter of a single corpuscle is the thirty-two-hundredth of an inch. After considerable practice, you will become accustomed to the apparent size of this object with a certain objective and eye-piece. This will aid in an approximate measurement of objects by comparison, and will further give the magnifying power of the microscope. If a corpuscle appears magnified to one inch in diameter, it is evident that the instrument magnifies thirty-two hundred times. Should the

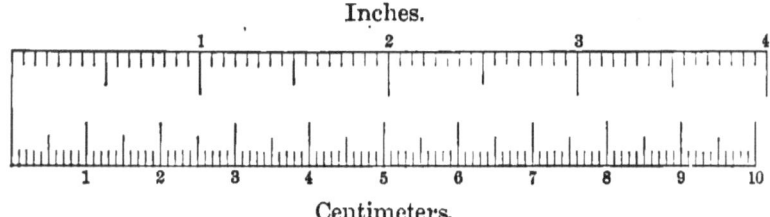

FIG. 3. ENGLISH AND METRIC SCALES.

diameter appear one-quarter of an inch, the power is eight hundred; one-eighth of an inch, four hundred, etc. The instrument which we have heretofore described, with the high-power in use and the tube withdrawn, will present the corpuscle as averaging very nearly one-eighth of an inch in diameter—× 400. While this gives a gross idea of amplification, the method will often prove to be inaccurate because of individual errors in the estimation of proportions.

Measurements made with the microscope are usually based on the metric system. The unit taken is one-thousandth part of a millimeter, or a micro-millimeter, called also a *micron*, for which the Greek letter μ has been taken as the symbol. Roughly, 1 μ equals $\frac{1}{25000}$ inch. A convenient instrument for measuring is an eye-piece micrometer, a ruled disk of glass, which may be placed within the eye-piece, and the diameters of objects read off by means of the ruled scale.

SKETCHING FROM THE MICROSCOPE

Let us most emphatically urge the practice of sketching in connection with microscopy. "I am no artist," or "I have no skill in drawing," is often the reply to our advice in this matter. We then suggest that no special skill is needed to begin with, only patience and a dogged determination to succeed. The pictures in the microscopic field have no perspective, and may be reproduced in outline merely. Begin with simple tissues, reserving intricate detail until a short period of practice gives the technique needed. We do not recommend the camera lucida, as our experience strongly impresses us with this as a fact, that he who cannot sketch without a camera will never sketch with one. Pencil drawings may be very effectively colored with our staining fluids, diluted if necessary.

PREPARATION OF TISSUES FOR MICROSCOPICAL PURPOSES

TISSUES ARE STUDIED BY TRANSMITTED LIGHT

The microscopical study of both normal and pathological tissues is invariably conducted by the aid of transmitted light.

Tissues, if not naturally of sufficient delicacy to transmit light, must in some way be made translucent.

Delicate tissues like omenta, desquamating epithelia, fluids containing morphological elements, certain fibers, etc., are sufficiently diaphanous, and require no preparation. Such objects are simply placed upon the glass slip, a drop of some liquid added, and, when protected by a thin covering glass, are ready for the stage of the microscope.

PREPARATION BY TEASING

The elements of structures mainly fibrous—*e.g.*, muscle, nerve, ligament, etc.—are well studied after a process of separation, by means of needles, known as teasing. A minute fragment of the organ or part, having been isolated by the knife or scissors, is placed upon a glass slip, and a drop of some fluid which will not alter the tissue added. Stout sewing-needles, stuck in slender wood handles, are commonly employed in the teasing process.

The separation of tissues is frequently facilitated by means of dissociating fluids, which remove the cement substance.

SECTION CUTTING

After having become familiar with the various elementary structures of animal tissues, we proceed to the study of their relation to organs.

As the teasing process is not available with such complicated structures as lung, liver, kidney, brain, etc., we resort to methods of slicing—*i.e., section cutting*.

Sections must be made of extreme tenuity, in order that the naturally opaque structures may be illuminated *by transmitted light*. This becomes an easy matter with such tissues as cartilage; but some, like bone, are much too hard to admit of cutting, and others are as much too soft; so that while certain tissues must be softened, the majority must be hardened. Fortunately, both of these conditions may be secured without in any way altering the appearance or relations of the structures. Hardening processes, from necessity, become a prominent feature in histological work; but we propose here to indicate some of the more useful methods of section cutting, reserving the hardening processes for another place.

FREE-HAND SECTION CUTTING

The students, when ready for this work, are provided with some tissue which has been previously hardened. We will take, for example, a piece of liver which has been rendered sufficiently firm for our work by immersion in alcohol, and proceed to direct the steps in obtaining suitable sections by the simple free-hand method.

We wish to strongly emphasize the importance of this mode of cutting. A moderate amount of practice will render the microscopist independent of all appliances, save those of the most simple character and which are always obtainable.

An ordinary razor with keen edge, and a shallow dish, preferably a saucer, partly filled with alcohol, are required. The razor best adapted to the work is concave on one side (the upper side, as seen in Fig. 4) and nearly flat on the other, although this is largely a matter of personal preference.

Fig. 4 indicates the proper position of the hands in commencing the cut. The beginner must follow directions closely until he acquires skill with practice. The student should be seated at a table of such height as to afford a convenient rest for the forearms. A small piece of tissue is held between the thumb and forefinger of the left hand, so that it projects slightly above both. (In the cut, a cube of tissue too small to handle in this way has been cemented to a cork with paraffin in the manner hereafter described, and the cork held as just mentioned.) The hand carrying the tissue is held over the saucer of alcohol. The razor, held lightly in the right hand, as seen in the figure, is, previous to

FIG. 4. FREE-HAND SECTION CUTTING.

making every cut, dipped flatwise into the alcohol so as to wet it thoroughly, and is then lifted horizontally, carrying several drops—perhaps half a drachm—of the fluid on the concave upper surface. The alcohol serves to prevent the section from adhering to the knife, and to moisten the tissue. If allowed to become dry, the latter would be ruined by alterations of structure.

Now, as to the manner of moving the knife. Resting the under surface upon the forefinger for steadiness, bring the edge of the blade nearest the heel to the margin of the tissue furthest from you. Then, entering the edge just below the upper surface of the tissue, with a light but steady hold draw the knife toward the right, at the same time advancing the edge toward the body. This passes the knife through the tissue diagonally, and leaves the upper surface of the latter perfectly flat or level. Remove the piece which has been cut, and repeat the operation. Do not attempt to cut large or very thin sections at first. A minute fragment, if thin, is valuable.

As the razor is drawn through the tissue, the section floats in the alcohol; depress the point of the knife, and the section will slide into the saucer of spirit, and thus prevent its injury. If it does not leave the knife readily, brush it along with a camel's-hair pencil which has been well wet with the alcohol.

Proceed in the above manner until the tissue is exhausted, when you will have a great number of sections, large and small,

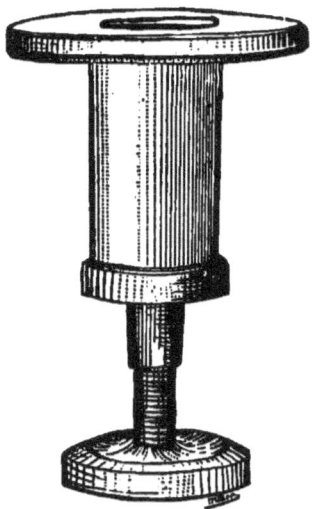

FIG. 5. STIRLING'S MICROTOME.

thick and thin. Selecting the thinnest, lift them carefully with a needle, one at a time, into a small, wide-mouthed bottle of alcohol; cork and label for future use.

When the work is finished, and before the spirit has evaporated from your fingers—it is impossible to avoid wetting the skin more or less—wash them thoroughly and wipe dry. This saves the roughening of the hands which is apt to result when alcohol has been allowed to dry upon them repeatedly.

SECTION CUTTING WITH THE STIRLING MICROTOME

Of the numerous mechanical aids to section cutting, we shall mention only two or three. One of the earlier and better known instruments is seen in Fig. 5. The Stirling microtome consists essentially of a short brass tube, into which the tissue is fixed either by pressure or by imbedding in wax. A screw enters below, which, acting on a plug, raises the contents of the tube. As the material to be cut is raised from time to time by the screw, it appears above the plate which surrounds the top of the tube. This plate steadies and guides the razor; and it is evident that more uniform sections may be cut with this little apparatus than would be possible with nothing to support the knife, or to regulate thickness, beyond the unaided skill of the operator.

Much depends upon the manner in which the material is fixed in the tube or well of the microtome. If the tissue be of a solid character, like liver, kidney, spleen, many tumors, etc., it may be surrounded with some carefully fitted pieces of elder-pith,* carrot,

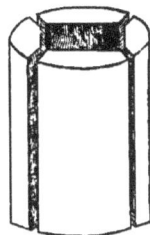

FIG. 6. MANNER OF CUTTING AND ARRANGING PIECES OF PITH, TURNIP, ETC., FOR SUPPORTING HARDENED TISSUE IN THE WELL OF A MICROTOME.

etc., and the whole pressed evenly and quite firmly into the well. A small piece of tissue which, by cutting, can be made somewhat cubical in shape, may be surrounded by slabs of pith, carrot, or turnip, shaped as in Fig. 6. Indeed, the fragments of imbedding material can be shaped so as to fit tissue of almost any form. Before the whole is pressed into the well of the microtome, the bottom, against which the brass plug fits, should be cut off square.

The wax method of imbedding is employed with tissues such as brain, lung, soft tumors, etc., which might be injured by the previous treatment. To three parts of paraffin wax (a paraffin candle answers perfectly) add one part of vaselin, and heat until thoroughly mixed. The microtome having been previously warmed —standing upright—is filled with the imbedding mixture. The

---

*The pith from the young shoots of *Ailantus glandulosus* (improperly called "Alanthus"), gathered in early autumn, is the best material for this method of imbedding with which I am acquainted. The wood is easily cut from the pith, and the latter is very large and firm.

piece of tissue is then carefully wiped dry with the blotting-paper, and, just as the imbedding begins to congeal around the edges, is pressed below the surface with a needle and held until the cool mixture fixes it in position. The whole is now allowed to become thoroughly cold. By turning the screw the plug of wax is raised;

FIG. 7. METHOD OF CUTTING SECTIONS WITH THE STIRLING MICROTOME.

and it must be gradually cut away, by sliding the knife across the plate, until the upper part of the tissue appears.

Before commencing to cut sections—however the tissue may have been imbedded—provide yourself with a saucer of alcohol and a camel's-hair pencil. Having wet the knife, turn the screw so that the tissue, with its imbedding, appears slightly above the

FIG. 8. THOMA MICROTOME.

plate of the microtome, and then, resting the blade of the razor on the plate (*vide* Fig. 7), make the cut precisely as in free-hand cutting. The section is then brushed off into the saucer, the screw turned up slightly, the razor wet, and a second cut made. These steps are repeated until the required number of sections has been obtained.

The imbedding material will separate from the cuts as they are floated in the alcohol. They may now be selected, lifted with the needle into clean spirit, and preserved, as before indicated, for future operations.

The finest section cutting can only be done with one of the more elaborate and expensive microtomes (Figs. 8 and 9). Each of these instruments consists of a heavy sliding knife-carrier, which moves on a level and with great precision, and of a device

FIG. 9. SCHANZE MICROTOME.

(screw or inclined plane) for elevating the object that is cut the desired distance after each excursion of the knife. The distance that the object is raised is, of course, the thickness of the section. Such microtomes are especially adapted to cutting frozen tissues or those imbedded in celloidin, but they may be used with paraffin imbedding. The Minot microtome, which is intended for cutting objects in paraffin, is preferable for the latter purpose, especially if a number of sections are to be kept in the order in which they were cut,—"serial sections."

SHARPENING KNIVES

In the majority of instances of failure to produce suitable sections for microscopical work, the cause can be set down to dull knives, and we would urge the student to practice honing until able to put cutting instruments in good condition. If he will but start properly, success is sure. Nine-tenths of the microtomes are purchased because of failure in free-hand work with a dull knife; but *no advantage will be gained by a machine, if the student be incapable of keeping the knife up to a proper degree of keenness.*

A knife is a wedge, and for our purpose the edge must be of more than microscopical tenuity—it being impossible with the microscope to discover notches and nicks if properly sharpened.

It is impossible to secure the best results with indifferent tools.

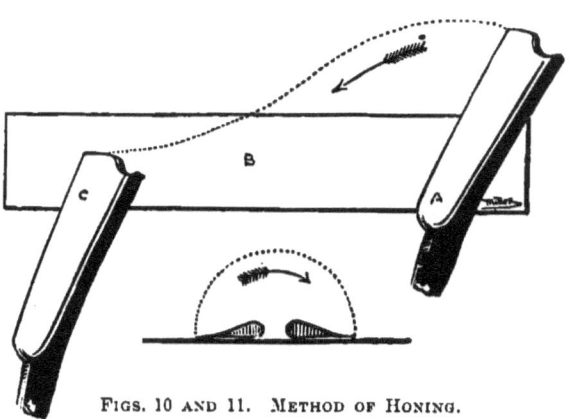

FIGS. 10 AND 11. METHOD OF HONING.

The knife is first brought with its heel in the position shown at A, Fig. 10. It is then drawn forward as indicated by the curved dotted line until, at the end of the stroke, the position C is attained. Fig. 11 indicates the method of turning the blade before reversing and between each stroke.

The knife should be hard enough to support an edge, but not so hard as to be brittle. The proper temper is about that given a good razor.

We need at least two hones—one comparatively coarse, for removing slight nicks, and another for finishing. The first part of the work is best done by means of a sort of artificial hone made of ground corundum. These are kept in stock by dealers in mechanics' supplies, of great variety in size and fineness. For razors a "00" corundum slip will best answer. This will very rapidly remove the inequalities from an exceedingly dull razor. A Turkish hone will be best for. finishing. For large knives we use a third very soft and fine stone.

Let the corundum slip be placed on a level support (or be fitted into a block like the carpenter's oil-stone), and cover the surface liberally with water.* The hones should always be worked wet. Place the knife flat on the stone near the right hand, as at A, Fig. 10. Draw steadily in the direction of the curved dotted line—*i.e.*, from right to left—holding the blade firmly on the stone, B, with slight pressure until the position C is attained. Rotate the razor on its back—*vide* Fig. 11—so as so bring the other side on the stone, and draw from left to right. Observe that as the knife is drawn from side to side (the edge invariably looking toward the draw) it is always worked from *heel to point*. The amount of pressure may be proportioned to the condition of the edge. If it be badly nicked, considerable pressure may be employed; while, as it approaches keenness, the pressure is to be lessened, until the weight of the blade alone gives sufficient friction.

Repeat the process fifteen or twenty times, and examine the blade. If the nicks are yet visible, continue honing until they can no longer be seen. Then draw the edge across the thumb-nail. Do this lightly, and the sense of touch will reveal indentions which the eye failed to recognize. Continue the use of the coarse stone until the edge is perfect, so far as the thumb-nail test indicates.

The knife is then to be carefully wiped, so as to remove any coarse particles of corundum, and applied to the wetted Turkish hone with precisely the same motions as were employed in the first process. After a dozen or two strokes, examine the edge by applying the palmar aspect of the thumb, with repeated light touches, from heel to point. This looks slightly dangerous to the novice, but it is an excellent method of determining the condition. Of course actual trial with a piece of hardened tissue is the best test.

A fine water-stone or the Belgian hone of the hardware shops may be used instead of the corundum hone.

It is best to finish with stropping, and often a knife may be sharpened by stropping alone. The leather of the strop should be glued to a support of wood to keep it flat. The movement is

---

*A few drops of glycerin added to the water retard evaporation, and appear to keep the surface of the hone in good condition.

the reverse of that employed for honing. The motion is from toe to heel, the back of the knife preceding the edge. Fig. 10, with the arrow reversed, illustrates the movement.

SUPPORTING TISSUES FOR CUTTING

Frequently small bits of tissue are required to be cut—pieces too small to be held with the fingers. We are in the habit of

FIG. 12. INSTRUMENT FOR SOLDERING TISSUE TO CORK SUPPORTS WITH PARAFFIN.

It consists of an awl-handle of wood into which a short piece of wire, preferably copper, is driven and bent as shown.

cementing such tissues into a hole in a bit of ailantus- or elder-pith, when the whole may be cut as one mass. Tissue is frequently cemented to cork for convenience of holding in free-hand

FIG. 13. METHOD OF CEMENTING TISSUE TO A CORK SUPPORT WITH PARAFFIN.

cutting; or the cork may be held in the vise of the microtome. The edge of the knife should not be allowed to touch the cork.

Fig. 12 shows a simple little instrument, very convenient for using paraffin as a cement. A piece of stout copper or brass wire is bent as indicated, pointed, and driven into an ordinary awl-handle. Paraffin wax possesses the very valuable property of remaining solid at ordinary temperatures, not cracking in the cold of winter nor softening in summer. It is unaltered by most reagents, is easily rendered fluid, and quickly solidifies. As a cement, it is invaluable to the microscopical technologist.

Fig. 13 indicates the method of cementing a piece of tissue to a cork or other support. The tissue having been properly placed, the wire tool is heated for a moment in the alcohol flame, and then

FIG. 14. CATHCART'S ETHER FREEZING MICROTOME.

touched to a cake of paraffin. The paraffin is melted in the vicinity of the hot wire, a drop adheres to the latter and is carried to the edge of the tissue. In the cut the wire tool is seen in the position necessary for cementing one edge. The wire being removed, the wax immediately cools and becomes solid. The other sides are afterward cemented in like manner. The whole is done in less time than is necessary to the description of the process.

THE FREEZING MICROTOME

After freezing a piece of tissue, sections may be cut without the delay required for imbedding in paraffin or celloidin. Fresh tissues may be used, or those hardened by some of the various fixing fluids, preferably formaldehyde, may be taken. Small pieces that have been hardened for a few hours in formaldehyde may be cut very thin after freezing, and this plan is most useful in pathological work. Sections of fresh tissues may be examined with the microscope directly, or they may be placed in formaldehyde, and afterwards stained and mounted as directed in the succeeding pages. Sections of hardened tissues may be stained and mounted in balsam.

Freezing of the specimen is done upon a metal box or plate, which may be attached to the ordinary microtome. Microtomes designed especially for freezing are also made (Fig. 14). Freezing is best accomplished by the escaping stream of carbon dioxide gas, derived from a cylinder of the compressed gas. An ether-spray may be used for the same purpose, or the water running from a salt and ice freezing mixture through a small metal case which may be placed on the microtome. The pieces of tissue should be not more than five millimeters thick.

FIXING OR HARDENING FLUIDS

We have already seen that most animal tissues are unsuitable for the production of thin sections until hardened.

It is also a fact, paradoxical though it may seem, that fresh tissues do not present truthful appearances of structural elements. The old-school histologists insisted upon the presentation of structures unaltered by chemical substances, while the modern worker has discarded such tissue with very few exceptions. Many descriptions for structure and growth, the result of study upon fresh material, have been proved by later methods grossly inaccurate.

It is impossible to remove tissues from the living animal and to subject them to microscopical observation without, at the same time, exposing them to such radical changes of environment as to produce structural alterations. Certain tissues presenting in the living condition stellate cells with the most delicate, though welldefined, branching processes, when removed from contact with the body, however expeditiously, afford no hint of anything resembling such elements, as they are quickly reduced to simple spherical outlines.

In short, it is impossible to study fresh material, as such, without constant danger of erroneous conclusions, as retrograde alterations of structure commence with surprising rapidity the moment a part is severed from the influences which control the complete organism.

From what has been said, we appreciate the necessity of agents which, when applied to portions freshly removed from the animal, or even before removal, shall instantly stop all physiological processes and retain the elements in permanent fixity.

Very much of the human structure which is available will be secured only after functional activities have long ceased, and the structure essentially altered. We are, therefore, compelled to resort to the use of material from the lower animals in very many instances.

ALCOHOL HARDENING

The tissue, whatever process may be in contemplation, having been removed from the body as quickly after death as possible, should, with a sharp scalpel, be divided into small pieces. Of the more solid organs, pieces not more than one-half to one centimeter thick will be sufficiently small, and they will harden rapidly. The smaller the pieces and the larger the quantity of hardening fluid, the more quickly will the process be completed. The volume of fluid should exceed that of the tissue at least twenty times. Wide-mouthed, well stoppered bottles, from one ounce to a pint, or even larger, are best, and they should be carefully labelled and kept in a cool place, with occasional agitation.

Quick Method.—A piece of any solid organ, say liver, spleen, pancreas, kidney, uterus, lymph-node, etc., not more than one-half centimeter thick, may be perfectly hardened in twelve hours by immersion in one ounce of ninety-five per cent. alcohol. No more should be thus prepared than is to be cut within twenty-four hours, on account of the shrinkage which results after the prolonged immersion of solid structures in strong spirit.

After the tissue has been one hour in the above, it may be hardened in one or two hours more if transferred to absolute alcohol. This method is of frequent advantage in pathological histology.

Ordinary Method.—Generally it is better to place the pieces of tissue first in eighty per cent. alcohol, and to transfer them after a few hours to ninety-five per cent. alcohol, in which the hardening is completed. In this manner the shrinkage and the extreme hardening of the surface, which result when strong alcohol alone is used, are avoided. The jar needs to be shaken occasionally.

The portions of tissue should be not more than one-half to one centimeter thick.

Alcohol is the most generally used of all hardening and fixing fluids. Tissues preserved in it will keep indefinitely, especially if the strong alcohol be replaced by eighty per cent. alcohol after a time. When bacteria are to be demonstrated in the organs, alcohol is as good as any agent that can be used. But when the blood, the nuclear figures, the elements of the nervous system, and the finer points in histological structure are to be studied, other fluids are more suitable.

MÜLLER'S FLUID

Bichromate of potassium . . . . 2.5 grams.
Sulphate of sodium . . . . 1 gram.
Water . . . . 100 c.c.

Müller's fluid is one of the most valuable of all fixing agents. The time required for hardening is six weeks or more, but hardening may be hastened by placing the jar in an incubator. The pieces of tissue must be small. The fluid must not become discolored, and must be changed frequently at first. It is most valuable for nervous tissues, which harden only after months. After hardening, the tissues must be washed in running water for twenty-four hours, and are then placed in alcohol, except nervous tissues, which are placed in strong alcohol without washing, when the Weigert hæmatoxylin stain is to be used. Müller's fluid preserves the blood-corpuscles in the organs admirably. The hardening can be hastened by adding ten per cent. of formaldehyde (forty per cent. solution). This mixture is known as ORTH'S FLUID. It has the advantages of Müller's fluid, while the hardening is completed in a week or less.

FORMALDEHYDE is a gas which is manufactured in a forty per cent. solution in water. The solution has a very pungent and irritating odor, and is a powerful germicide. It is most valuable for preserving large anatomical and pathological specimens, which keep their natural colors in it much better than in alcohol. It is also useful in histological work as a fixing agent. Pieces of tissue one centimeter thick are hardened in twenty-four hours in ten parts of the forty per cent. formaldehyde solution of commerce, and ninety parts of water. Very small pieces of tissue may be hardened in formaldehyde for a few hours, and excellent sections may then be cut directly after freezing. The staining properties of tissues hardened in this fluid are well preserved. The specimens may be kept in it indefinitely.

PICRIC ALCOHOL (GAGE)

Picric acid	2 grams.
Alcohol	500 c.c.
Water	500 c.c.

This is an excellent fixing agent for most tissues. Hardening is completed in one to three days.

OSMIC ACID is sold in hermetically sealed capsules. It is used in one per cent. solution in distilled water. The solution should be freshly made when possible. It deteriorates in the light, and must be kept in a dark closet. Its vapor is very irritating. The usefulness of osmic acid depends largely upon its property of staining fat black. It is often employed in pathology. In histology it is most valuable for nervous tissues. It stains the medullary sheaths of medullated nerve-fibers black. The pieces of tissue must be not more than four millimeters thick, as osmic acid has very feeble penetrating power. They are left in the solution twenty-four hours, washed thoroughly, and transferred to eighty per cent. alcohol. In preparing specimens for microscopical study, it is best to avoid using turpentine or xylol, which dissolve the stained fat, if that is to be preserved.

FLEMMING'S SOLUTION

Two per cent. osmic acid solution	4 parts.
One per cent. chromic acid solution	15 parts.
Glacial acetic acid	1 part.

The pieces of tissue should be four millimeters or less in thickness. They are left in the fluid twenty-four hours, and, after thorough washing, are transferred to alcohol. This fluid is designed to preserve the karyokinetic figures, using safranin or hæmatoxylin to stain them. It is also admirable to fix cellular structures in general, and it stains fat black as well as osmic acid alone.

Another formula proposed by Flemming for fixing the karyokinetic figures consists of

Chromic acid	.2 grams.
Glacial acetic acid	.1 c.c.
Water	100.0 c.c.

It serves very well all the purposes of hardening agents. Using small pieces, leave them in the solution twenty-four hours. After washing, place in eighty per cent. alcohol, and subsequently in strong alcohol. Stain with safranin or hæmatoxylin.

CHROMIC ACID FIXING AND HARDENING

Chromic acid is a very deliquescent salt, and is best preserved by making a strong solution at once, and then diluting it as may be needed. A stock solution may be made as follows:

    Chromic acid (crystals) . . . . 25 grams.
    Water . . . . 75 c.c.

For general use, dilute three parts with six hundred parts of water, which gives a strength of nearly one-sixth of one per cent.

The tissue, as soon as secured and properly divided, is placed in the above, remembering the rule regarding quantity. Change in twenty-four hours to fresh solution, and again on the third day. In seven days, or thereabout, change the fluid again. The tissue must now be watched carefully; and when, on cutting through a piece, the fluid is found to have stained the blocks completely, taking from two to three, or even four weeks, remove to a large jar of clear water and wash, preferably with running water, for twenty-four hours. The washing having removed the chromic acid, the tissue is further hardened in alcohol.

Very small pieces of tissue may be hardened in one or two days. Chromic acid is useful to preserve the nuclear figures.

ERLICKI'S FLUID

    Bichromate of potassium . . . . .   25 grams.
    Sulphate of copper   10   "
    Water . . . . 1,000 c.c.

This may be employed in precisely the same manner as the dilute chromic acid solution.

DECALCIFYING PROCESS

    Six per cent. chromic acid solution . . . . . . .9 parts.
    Nitric acid, C. P. . . . . 1 part.
    Water . . . . 90 parts.

The earthy salts may be removed from teeth and small pieces of bone with a liberal supply of the above in about twenty days. A frequent change of the solution will greatly facilitate the process, and an occasional addition of a few drops of the nitric acid may be made with very dense bone. After the removal of the lime salts, the pieces may be preserved in alcohol until such time as sections are needed, when they may be cut with the microtome without injury to the knife.

DISSOCIATING PROCESS (W. STIRLING)

Artificial Gastric Fluid

    Pepsin . . . . 1 gram.
    Hydrochloric acid . . . . 1 c.c.
    Water . . . . 500 c.c.

This process depends for its value upon the fact that certain connective tissues are more rapidly dissolved by the fluid than others.

IMBEDDING

The best and thinnest sections can only be cut when properly hardened tissues are used, and when the meshes of the tissue have been filled with some substance which supports the delicate elements. The substances used for this purpose are paraffin and celloidin, or collodion.

THE PARAFFIN METHOD

a. Pieces of tissue 2–3 mm. thick are to be placed in ninety-five per cent. alcohol for twenty-four hours.

b. In pure chloroform six to eight hours.

c. In a saturated solution of paraffin in choloroform two to three hours.

d. In melted paraffin, having a melting point of 50° C., which requires the use of a water bath or oven, one to six hours. The chloroform must be entirely driven off, and the tissue thoroughly infiltrated.

e. Change to fresh paraffin for a few minutes.

f. Finally place the tissue in a small paper box and pour the melted paraffin about it. Harden as quickly as possible with running water. It is important to fix the piece of tissue in a suitable position, if the position is of importance, before pouring in the melted paraffin.

Sections of exquisite thinness may now be cut. The knife need not be wet. Paraffin imbedding is especially adapted to making serial sections.

In order to mount the sections, proceed as follows:

a. Place the sections on a slide. Add a thin solution of gum arabic, upon which they float. Warm slightly, when the sections will flatten nicely. Drain off the superfluous gum solution, leaving the sections in their proper positions. Let them dry for some hours, and they will be firmly fastened to the slide.

b. Dissolve out the paraffin in one of the numerous solvents (xylol, half an hour or less).

c. At this point, unless the piece of tissue was stained in bulk before imbedding, the xylol should be washed off with alcohol and

d. The section stained with one of the dyes described hereafter.

e. Dehydrate in alcohol.

f. Clear in some suitable agent, as xylol or oil of cloves.

g. Mount in balsam.

CELLOIDIN INFILTRATION

Certain structures require permanent support—*i.e.*, not only while being cut, but during the subsequent handling of the sections. The celloidin infiltrating process is best adapted to such material. Considerable time is needed for the successful employment of the process, but results can be secured that cannot be equaled with any other method.

Celloidin is the proprietary name of a sort of pyroxylin, very soluble in a mixture of ether and alcohol, producing a *collodion*. If thick collodion be exposed for a few moments to the air it becomes semi-solid—not unlike boiled egg-albumen; and to this property is due the value of a solution of celloidin in histology. It may be used as follows:

To a mixture of equal parts of ether and alcohol add celloidin* until the thickest possible solution has been obtained.

A piece of alcohol-hardened tissue, having been selected and kept for the preceding twenty-four hours in a mixture of equal parts of alcohol and ether, is placed in about an ounce of the solu-

---

*We find, after repeated trial, that the ordinary soluble gun-cotton, such as is employed by photographers, is in no way inferior to the celloidin.

tion and allowed to remain twenty-four hours. The bottle containing the whole should be well corked, to prevent evaporation.

The tissue after infiltration is to be placed on a wooden block, and allowed to remain in the open air for a few minutes, after which it should be plunged into a mixture of alcohol two parts, water one part. Here it may remain for twenty-four hours, or until wanted.

Cut in the usual way, using a mixture of alcohol two parts, water one part, for flooding the knife. The section should be finally preserved in the same instead of pure alcohol, which would dissolve the celloidin.

In infiltrating the tissue with the collodion it is best, especially if it be very dense in parts, to use first a thin and subsequently the thick solution. A more perfect infiltration is often obtained in this way. In some cases we have been obliged to continue the maceration for several days. The solution should be kept in well stoppered bottles, as the ether is exceedingly volatile. Should the collodion at any time become solid from evaporation, it may be easily dissolved by adding the ether and alcohol mixture.

The process is of inestimable value where delicate parts are weakly supported, and where it is important to preserve the normal relations. The gelatin-like collodion permeates every space, and as it is not to be removed in the future handling of the sections, it affords a support to portions that would otherwise be lost or distorted. It offers no obstruction to the light, being perfectly translucent and nearly colorless.

STAINING AGENTS AND METHODS

STAINING FLUIDS

It is a very interesting fact (and one upon which our present knowledge of histology largely depends) that, on examination of tissues which have been dyed with special colored fluids, the dye will be found to have colored certain anatomical elements very deeply, others slightly, while others still remain unstained.

Certain dyes are called *general* or *ground* stains, because they stain all parts of a tissue alike, or nearly so. Others, which are entitled *selective*, exhibit an affinity for some particular structure, usually the nucleus of the cell. Hæmatoxylin, or logwood, for instance, has such an affinity for nuclei. The whole nucleus is not stained, but certain threads in it, which ordinarily appear as granules. This part of the nucleus is called chromatin, on account of its affinity for dyes. In a tissue colored with hæmatoxylin, the minute granules of the nuclei are so deeply stained in the logwood dye as to appear almost black. The nuclei are plainly stained, while the limiting membrane of cells is usually but slightly colored. Old, dense connective tissues stain feebly, or fail entirely to take color. The differentiation is, without doubt, due to chemical action between the elements of the dye and those of the tissue.

A very great number and variety of materials have been used for histological differentiation, and a simple enumeration of them all would very nearly fill the remainder of our pages. It will be found, however, that leading histologists confine themselves to two or three standard formulæ for general work. We shall notice only those methods that have been thoroughly demonstrated by years of employment as best for the purpose suggested. Special cases will require special treatment, which will be indicated in proper connection.

It is important in all cases to secure the purest and most concentrated dyes obtainable. It is better to make your own solutions than to buy them already prepared.

DELAFIELD'S HÆMATOXYLIN

Hæmatoxylin crystals . . . . 4 grams.
Alcohol . . . . 25 c.c.
Ammonia alum . . . . 50 grams.
Water . . . . 400 c.c.
Glycerin . . . . 100 c.c.
Methyl alcohol . . . . 100 c.c.

Dissolve the hæmatoxylin in the alcohol, and the ammonia alum in the water. Mix the two solutions. Let the mixture stand four or five days uncovered; it should have become a deep purple. Filter and add the glycerin and the methyl alcohol. After it has become dark enough filter again. Keep it a month or longer before using; the solution improves with age. At the time of using, filter and dilute with water as desired.

EOSIN

Eosin is an aniline dye, sold in the form of a red powder. It is best to keep on hand a saturated solution in alcohol. A few drops of this stock solution may be added to a small dish full of water at the time of using.

BORAX-CARMINE (GRENACHER)

Carmine	2.5 grams.
Borax	4.0 grams.
Alcohol (70%)	100.0 c.c.
Water	100.0 c.c.

Rub the carmine and borax together. Dissolve them in the water, which should be hot. The alcohol may be added when the mixture is cold. The stain will be available after two weeks, but improves with age. This solution is especially suitable for the staining of whole masses of tissue before imbedding, i.e., in bulk. For staining in bulk, leave the tissue in the carmine for twenty-four hours. It may be used for sections, however, which are to be left in the stain fifteen minutes or longer. In all cases carmine staining is to be finished with the use of acid alcohol, which differentiates the elements.

ACID ALCOHOL

Alcohol (70%)	100 c.c.
Hydrochloric acid (strong)	1 c.c.

Sections stained in carmine are placed in acid alcohol for a few minutes. They acquire a brilliant scarlet color. For specimens stained in bulk, dilute the acid alcohol with twice as much 70 per cent. alcohol, and leave the tissue in the mixture twenty-four hours.

PICRO-CARMINE

Carmine	1 gram.
Aqua ammonia (strong)	5 c.c.
Water	50 c.c.
Saturated watery solution of picric acid	50 c.c.

Add the picric acid solution after dissolving the carmine in the diluted ammonia. Let the mixture stand uncorked for two days, and filter. The carmine gives a nuclear stain, while the picric acid serves as a counter stain.

IODINE SOLUTION

Iodine	1 gram.
Potassium iodide	2 grams.
Water	300 c.c.

WEIGERT-PAL HÆMATOXYLIN METHOD

This stain is employed for nervous tissues containing medullated nerve-fibers. The medullary sheaths of these fibers are stained intensely black, while the other elements of the tissue remain pale. It is used chiefly for the spinal cord and the brain.

The tissue is first hardened in Müller's fluid. Unless the pieces of tissue are very small, the hardening requires months. The fluid must be changed frequently. Hardening is completed in strong alcohol, without washing in water. Imbed in celloidin.

The sections are overstained with hæmatoxylin, and subsequently are partly decolorized. The medullary sheaths retain the stain with greater tenacity than the other elements of the tissue. The solutions used are as follows:

> Hæmatoxylin crystals . . . . 1 gram.
> Alcohol . . . . 10 c.c.
> Water . . . . 90 c.c.

Boil and filter. Allow the solution to stand a week or two. If used sooner, add a drop of saturated solution of lithium carbonate to a part of the stain in a small dish.

> Permanganate of potassium . . . . 1 gram.
> Water . . . . 400 c.c.
> One per cent. solution of oxalic acid.
> One per cent. solution of sodium or potassium sulphite.

Mix the oxalic acid and sodium sulphite in equal parts at the time of using. Sulphurous acid is formed.

The steps in staining are as follows:

a. Sections are placed in the hæmatoxylin for twenty-four hours, and are intensely stained, becoming nearly black. The process may be hastened by letting them stand in an incubator, where they may be sufficiently stained in a few hours.

b. Wash in water.

c. Place the sections in the permanganate of potassium until they acquire a dark brown color (one-half to five minutes).

d. Wash in water.

e. Place the sections in the mixture of oxalic acid and sodium sulphite. The brown color should give way to a blue black in the white matter, while the gray matter becomes nearly white. If the sections remain too dark they may be returned to the permanganate for a few minutes, and then to the sulphurous acid again, always washing in water between the two changes.

f. Wash thoroughly in water, and mount in the usual way in balsam.

ANILINE DYES

The substances known as aniline dyes are derivatives of coal tar, but not always of aniline. These dyes have become of great importance in all kinds of biological work. The number of different compounds is very large, and only a few of the most common can be mentioned. None but the purest dyes should be used, and those manufactured by Grübler can be recommended. They are sold in the form of powders. American agents for Grübler's dyes are Eimer & Amend, of New York city. It is simplest to classify the dyes as basic or acid. Fuchsin, methylene blue, gentian violet, and safranin are basic dyes. They have an affinity for nuclei and for bacteria. Eosin, picric acid, and acid fuchsin are acid dyes, tending to stain tissues diffusely. The use of eosin and picric acid has been described on pages 28 and 29.

Certain cells have granules in their protoplasm. The granules of some cells manifest an affinity for basic dyes (basophile, δ and γ), others for acid dyes (acidophile or eosinophile, α), and others for a mixture of the two (neutrophile, ϵ), and others still for both the acid and the basic (amphophile, β).

It is best to keep on hand saturated alcoholic solutions of these dyes, from which watery solutions may be made when needed, by adding a few drops of the alcoholic solution to a small dish filled with water.

The basic dyes may be used as nuclear stains as follows:

a. Stain sections in a strong watery solution of the dye five minutes or more. The sections will be somewhat overstained.

b. Wash in one per cent. acetic acid a few seconds.

c. Alcohol. Dehydration must be done rapidly, as alcohol extracts the color from the tissues. It must, nevertheless, be thorough, as xylol, which is used in the next step, only mixes with strong alcohol.

d. Xylol. This agent is the one used to clear specimens after staining with basic aniline dyes, because most of the oils slowly dissolve out the aniline colors.

e. Balsam.

Among the dyes used in this manner, SAFRANIN is to be especially recommended as a nuclear stain. The sections are treated according to the programme just given, except that they should remain in the safranin solution (one per cent.) some hours. Safranin has been employed largely in staining karyokinetic figures, after hardening with Flemming's solution (page 23).

NIGROSIN is an aniline color used mostly for nervous tissues. It is valuable when results are desired at once, without waiting for the tedious hardening process required for the Weigert-Pal method.

a. Stain sections in strong watery solution of nigrosin five to ten minutes.

b. Wash and mount in the usual way in balsam, clearing in xylol.

VAN GIESON'S STAIN

Acid fuchsin (one per cent. watery solution). . . . 15 c.c.
Picric acid (saturated watery solution) . . . . . . 50 c.c.
Water. . . . . 50 c.c.

a. Stain sections in hæmatoxylin.

b. Wash in water.

c. Stain in the picric acid and acid fuchsin from three to five minutes.

d. Wash in water and mount in the usual way in balsam.

The Van Gieson stain is used chiefly for connective and nervous tissues.

EHRLICH TRICOLOR STAIN

Saturated watery solution orange G. . . . . . 120–135 c.c.
    "   "   "   acid fuchsin . . . 80–165 c.c.
    "   "   "   methyl green . . .   125 c.c.
Glycerin . . . . 100 c.c.
Absolute alcohol . . . . 200 c.c.
Distilled water . . . . 300 c.c.

This formula is only one of a number with which Ehrlich's name is associated, as well as those of Biondi and Heidenhain. A powder containing the dyes already mixed is sold by dealers, and usually works very well. It may be used to stain sections of tissues, but is employed mostly with preparations of blood, dried on cover-glasses and fixed by heating. Stain five minutes or less. It is designed to stain the neutrophile granules of certain leucocytes, which become colored reddish brown. Eosinophile granules become brilliant red. Nuclei are stained green by it, while the red corpuscles are orange-red or brown. Its use will be referred to in the chapter on blood.

METALLIC STAINS OR IMPREGNATIONS

Compounds of silver, gold, mercury, and osmium are used to color particular elements in the tissue, or they are precipitated in certain structures as opaque deposits. Osmic acid has already been described on page 23. Of the other substances, NITRATE OF SILVER is the most important.

This salt is used in dilute solutions (1–300, 1–500) made with distilled water. It is most valuable to stain the cement-substance between the endothelial cells of membranes, like the peritoneum. The membrane is stained while fresh, and must be washed free of all albuminous substances which might precipitate the silver. It should also be stretched over the rim of a dish. The solution of nitrate of silver should be allowed to act upon the membrane, care being taken to have it reach all parts of the surface. It has little penetrating power. After five to ten minutes wash in water, and expose to the sunlight, in which the color becomes brown, owing to dark lines which develop between the cells. Mount in glycerin or balsam.

GOLGI METHOD FOR STAINING BRAIN AND SPINAL CORD

There are many formulas which different investigators have used with more or less success. All of them seek to impregnate the tissue with silver or mercury salts, which become precipitated on some of the nerve-elements and render them visible. At best, the Golgi method is uncertain, but when it is successful, the preparations obtained are of great beauty and value.

The following procedure can be recommended:

a. Small pieces of fresh tissue are hardened from two days to a week in

> Osmic acid, one per cent. . . . . 10 c.c.
> Potassium bichromate, three and one-half per cent. 40 c.c.

The time required varies according to the part of the tissue to be stained. Neuroglia requires the shortest and nerve-fibers the longest time. For ganglion-cells the time should be intermediate between these.

b. The tissue is placed in a three-fourths per cent. solution of nitrate of silver for one to six days.

c. Cut rather thick sections, about 50 μ. Alcohol is to be used as little as possible. Sections may be cut free-hand or between pieces of elder-pith, or by fastening on the block with paraffin (page 18); or by imbedding rapidly in celloidin, taking about twenty minutes for all the steps.

d. Alcohol, xylol, and balsam, as usual.

e. Mount without a cover-glass, and keep in the dark.

COX'S MODIFICATION OF THE GOLGI METHOD

Golgi found that the bichloride of mercury could be used for the impregnation of nervous tissues in much the same manner as the nitrate of silver. The formula proposed by Cox has been highly recomended.

    Bichromate of potassium (five per. cent. solution) 20 parts.
    Bichloride of mercury (five per cent. solution) . . 20 parts.
    Simple chromate of potassium (five per cent. sol.) 16 parts.
    Distilled water . . . . 30 to 40 parts.

Specimens should be left in the mixture for one month in summer, and for two to three months in winter. The impregnation should take place uniformly throughout the preparation.

INJECTION OF BLOOD-VESSELS

To make the small blood-vessels appear in microscopical preparations they may be filled with some colored substance. An entire animal recently killed, or merely one organ, may be injected. A canula is tied in the proper vessel and injected with a syringe, or from a flask holding the coloring substance, which is emptied by mercurial or water pressure. Among many formulas for injecting fluids, the following may be used:

    Soluble Berlin blue . . . . 3 grams.
    Water . . . . 600 c.c.

This mixture has the advantage that it may be used cold.

CARMINE-GELATIN must be kept warm while injection is proceeding, as must the object which is being injected. Two and one-half grams of carmine are rubbed up with a little water, and strong ammonia added, a drop at a time, till the carmine is dissolved. Then filter. Five grams of gelatin having previously been dissolved in water, add the carmine solution. Now neutralize the carmine-gelatin exactly with acetic acid. As the neutral point is approached, the acid must be diluted and added cautiously; filter.

CLEARING AGENTS

The commonest method of preparing sections is to mount them finally in Canada balsam. Staining is usually performed in watery solutions. Water does not mix with balsam. After staining, therefore, the water is to be removed with alcohol, which has an affinity for water. The alcohol must in turn be removed with some substance which will mix with balsam. The various fluids used for clearing have this property, and also make the tissue transparent. ANILINE OIL is a clearing agent which will itself remove water from the tissues without the use of alcohol. It does not dissolve celloidin. It extracts the aniline colors, however. XYLOL can only be used when dehydration is perfect. It does not dissolve celloidin, nor extract the aniline dyes. It is often used mixed with one-third of its volume of carbolic acid—CARBOL-XYLOL. When an agent that will not dissolve celloidin is desired, a cheap and excellent substitute for the last is the CARBOL-TURPENTINE of Gage:

```
Carbolic acid crystals (melted) . . . . . . . . . . 40 c.c.
Oil of turpentine . . . . . . . . . . . . . . . . . 60 c.c.
```

The essential oils and CREOSOTE are used frequently for clearing, for example, the OILS of ORIGANUM, of BERGAMOT, and of THYME. OIL OF CLOVES is used very extensively, but it has the property of removing the aniline dyes quite rapidly, and it dissolves celloidin. It may be used to clear celloidin sections, except the most delicate, if care be used. Delicate sections should be cleared on the slide. Other sections may be placed in a small dish of the clearing fluid.

MOUNTING MEDIA

CANADA BALSAM is the medium most used for the permanent preservation of microscopical preparations. It should be dissolved in xylol, which does not affect the aniline stains.

DAMMAR is sometimes used in the same manner as balsam. Balsam and dammar may be kept in large-mouthed bottles, from which they are removed with glass rods. They are also sold in flexible lead tubes, which make a convenient way of handling them.

Fresh tissues are often studied in a six-tenths of one per cent. solution of sodium chloride in water—NORMAL SALT SOLUTION.

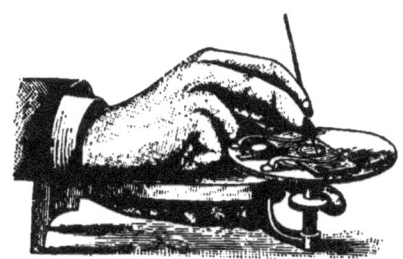

FIG. 15. USING TURN-TABLE—AFTER FREEBORN.

Objects are often mounted in GLYCERIN, especially those that would be injured by alcohol. It is best to use a circular cover-glass, and to surround the edge with some soluble cement, using a *turn-table* for this purpose.

Zinc cement, asphalt varnish, and other suitable cements may be purchased from dealers in microscopical supplies.

STAINING METHODS

HÆMATOXYLIN STAINING PROCESS

You will require for future work a needle like Fig. 16, several saucers, preferably of white ware; a few watch-glasses (large, odd sizes are usually cheaply obtainable at a jeweler's); half a dozen

FIG. 16. NEEDLE FOR LIFTING SECTIONS, ETC.

glass saltcellars—small ones known as "individual salts,"—and a two-ounce, shallow, covered porcelain box, such as druggists use for ointments, dentifrices, etc.

Place on the work-table (best located so as to be lighted from your side and not from the front) in order, as in Fig. 17:

1. *Watch-glass*, containing say 10 c.c. diluted hæmatoxylin.
2. *Saucer*, filled with water.
3. *Saltcellar*, filled with alcohol.
4. The *covered porcelain box*, containing about 20 c.c. oil of cloves* or carbol-turpentine.

Select a section from some one of your stock bottles, lifting it out with the needle, and place it in the hæmatoxylin solution. The

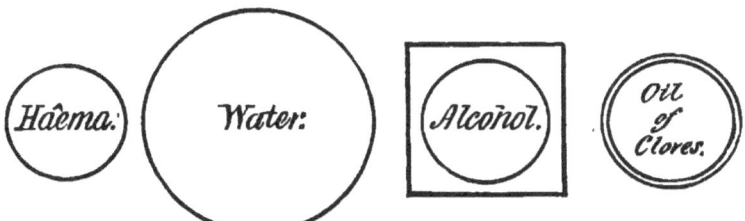

FIG. 17. DIAGRAM INDICATING THE SUCCESSIVE STEPS IN STAINING WITH THE HÆMATOXYLIN SOLUTION.

section, having been taken from alcohol and transferred to a watery staining fluid, will twirl about on the surface of the latter, inasmuch as currents are formed by the union of the water and the spirit.

"How long shall I let the section remain in the hæmatoxylin?" The only answer we can give is, "Until properly stained." Nothing but experience will give you any more definite information. Much depends upon some peculiar property in the tissues: some stain rapidly, others stain very slowly. The strength of the dye is another determining factor. Usually with the hæmatoxylin formula, as given, from two to three minutes will suffice.

Place the needle under the section (if the fluid be so opaque as to hide the tissue, place the watch-glass over a piece of white paper or a bit of mirror) and gently lift it out; drain off the adhering drop of dye on the edge of the glass, and drop into the saucer of water. Here we can judge as to the color, and we, perhaps, find it to be of a light purple—too light, so you may return it to the hæmatoxylin for another period of two or three minutes, which will probably give sufficient depth.

As the section floats on the washing water, you will notice that the latter will be colored by the dye, some of which leaves the

---
*Although oil of cloves is the clearing agent mentioned throughout this book, it is to be remembered that it dissolves celloidin, and that carbol-turpentine is preferable for sections if they are delicate.

tissue. Allow the water to act until no more color comes out. The tint of the section changes from purple to violet, and the water must be allowed to act until the change is complete.

If you were to examine your section at this stage, you would find it opaque, and as we are obliged to study our objects mainly by transmitted light, we must find some means of securing translucency. The essential oils are used for this purpose, oil of cloves being commonly employed. Lift the section from the water with the needle, let it drain a moment, and then drop it into the alcohol with which the saltcellar was filled. The object of this bath is the removal of the water from the tissue, and this will be accomplished in from five to ten minutes. Again lift the section and place it in the oil of cloves. The tissue floats out flat, and in a few minutes sinks in the oil.

We might proceed to the examination of the stained section; but we shall ask you to let it remain in the oil, covering the box carefully to exclude our great enemy, dust, until we have learned more about staining.

To recapitulate: The essential steps in the hæmatoxylin process are:

1. Staining the tissue—hæmatoxylin.
2. Washing—water.
3. Dehydrating—alcohol.
4. Rendering transparent—oil of cloves.

As the section is put in the dye, care should be taken to so float it out that it may not be curled. This is easily done with the needle. After the alcohol bath, however, this becomes difficult, as the tissue is rendered stiff by the removal of the water.

This is the simplest and best of all methods for general work, and you are advised to master every detail of the process. After reading the directions which we have given, and having never seen the work actually done, it will not be singular if you conclude that the staining of tissues is a tedious and slow process; but after a month's work you will be able to stain fifty different sections in half an hour, and have them ready for mounting.

HÆMATOXYLIN AND EOSIN—DOUBLE STAINING

Very beautiful and valuable results in differentiation are obtained by staining first with hæmatoxylin and subsequently with eosin. Eosin, a coal-tar derivative, stains most animal tissues pink, and it affords with the hæmatoxylin a good contrasting color. The tissue is to be stained in hæmatoxylin and washed in water as usual; then it is placed in the eosin solution, and afterward washed again. The subsequent treatment is as with the plain hæmatoxylin process; viz., dehydration with alcohol, after which the oil of cloves.

The diagram, Fig. 18, shows the process complete:
1. *Watch-glass* with hæmatoxylin.
2. *Saucer* with water.
3. *Watch-glass* two-thirds filled with water, with five drops of eosin solution added.
4. *Saucer* containing water.
5. *Saltcellar* filled with alcohol.
6. *Covered oil-dish*.

The eosin stains very quickly, generally in about a minute. Care should be taken not to overstain with it, as it cannot be

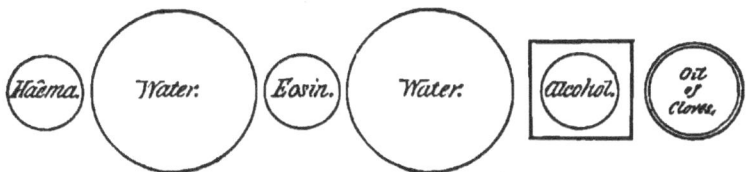

FIG. 18. DIAGRAM INDICATING THE SUCCESSIVE STEPS IN DOUBLE STAINING WITH HÆMATOXYLIN AND EOSIN.

washed out. If the sections are found at any time to be overstained with hæmatoxylin the color may be removed to any desired extent by floating them in a weak solution of acetic acid. They must afterward be washed very thoroughly in clean water.

BORAX-CARMINE STAINING PROCESS

Arrange your materials as in the diagram, Fig. 19.
1. *Watch-glass* two-thirds filled with the carmine fluid.
2. *Saucer* containing about an ounce of alcohol.
3. *Saltcellar* filled with alcohol containing one per cent. of hydrochloric acid.
4. *Saltcellar* with alcohol.
5. *Porcelain dish* containing oil of cloves.

The carmine solution will stain ordinarily in fifteen or twenty minutes. After the section has been in the dye for a few minutes, lift it with the needle, drain, and transfer to the saucer containing alcohol. You will then be enabled to determine whether the section is sufficiently stained; it should be a deep, opaque red. The alcohol washes off the section, removing the adhering solution of carmine.

The carmine must now be fixed in the tissue, or *mordanted;* and this you proceed to do by transferring the section to the watch-

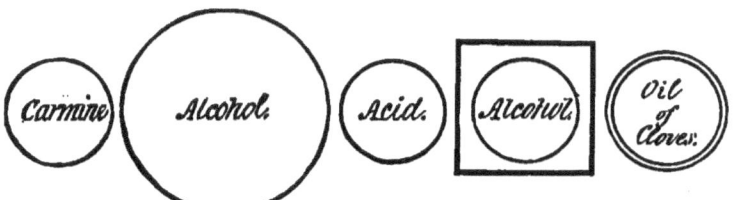

FIG. 19. DIAGRAM INDICATING THE SUCCESSIVE STEPS IN STAINING WITH BORAX-CARMINE.

glass containing acid alcohol. Notice the change in color, from a dull red to a bright crimson, and when the change is complete, lift it into the saltcellar containing clean alcohol. This dissolves out the acid. Five minutes suffice for this washing, after which transfer to the oil of cloves.

This process does not give as sharp contrasts as the hæmatoxylin and eosin, but it is simpler and very permanent. It is best to select some one process for general work, *and adhere to it.* The acid of the carmine process must be guarded with extreme care, as the smallest particle is sufficient to spoil the hæmatoxylin solution. Look to it that the dishes are kept scrupulously clean, and the same care must be bestowed upon the needles, forceps, and all other instruments.

Picro-carmine may be used instead of borax-carmine. The sections may need to remain in picro-carmine as much as an hour to become thoroughly stained. Acid alcohol is not to be employed after staining. In dehydrating, it is well to add some picric acid to the alcohol, in order to prevent extracting the yellow stain from the specimen.

You may, of course, stain several sections at once, providing you take care to keep them from rolling up or sticking together.

While the vessels which we have recommended will be found of convenient, proportionate, and economical size for general work, larger ones are sometimes needed; and almost any glass or porcelain vessel may be impressed for duty.

MOUNTING OBJECTS

CLEANING SLIDES AND COVERS

When purchasing slides, let us urge you to get them of good quality. The regular size is one by three inches, and the edges should be smoothed. As furnished by the dealers they are usually quite clean, and only require rubbing with a piece of old linen to prepare them for use.

The cover-glasses should be thin, not over $\frac{1}{100}$ of an inch, called in the trade-lists "No. 1." Circles or squares three-quarters of an inch in diameter are generally convenient. They must be thoroughly cleaned. Drop them singly into a saucer containing hydrochloric acid. Then pour off the acid, and let clean water run into the dish for several minutes. Drain off the water and pour a little alcohol on the covers. Remove them one at a time with the forceps or needle, and wipe dry with old linen.* The glass may be held between the thumb and forefinger, the linen being interposed. Very slight pressure and rubbing will complete the process. The surface of the glasses should be brilliant, and they should be preserved for future use in a dust-tight box.

TRANSFERRING THE SECTIONS TO THE SLIDE

The section is to be taken from the oil with a section-lifter or spatula, Fig. 20. Smooth, stiff paper, cut in strips, may be used in the same way.

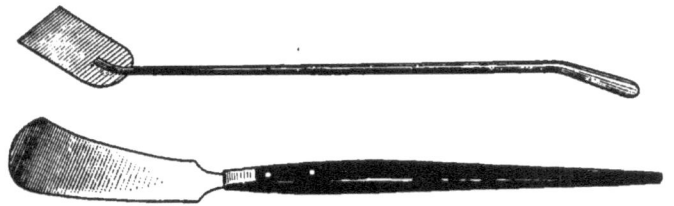

FIG. 20. SECTION-LIFTERS.

In changing the sections on the needle or spatula from one fluid to another, as from alcohol to oil of cloves, it is well to touch the

---

*We are indebted to Professor Gage, of Cornell University, for suggesting the use of Japanese tissue-paper for wiping cover-glasses, lenses, etc. Ordinary manilla toilet-paper is also an excellent material for such work.

edge of the section to a sheet of blotting-paper or filter-paper, to remove as much as possible of the first fluid and prevent its diluting the second.

Place a clean slide on the table before you, and, with the section-lifter used like a spoon, dip up one of the sections from the clove oil. By inclining the lifter, the section may be made to float to

FIG. 21. METHOD OF LABELING A MOUNTED SPECIMEN.

the center of the slide. A small sable brush is often convenient for coaxing the section off the lifter.

If it were our present object to simply examine the section, we could drop a thin cover-glass on the specimen, and it would be ready for study. Such an object would afford every requirement for present observation, but would not be permanent. The oil of cloves would evaporate after a few days, and the section be ruined. We proceed to make a permanent mounting of our object.

The clove oil surrounding the section on the slide is first to be removed, and it can easily be done by means of blotting-paper. With a narrow slip of thin filter-paper wipe up the oil, exercising care not to touch the section, or it will become torn. Proceed carefully, taking fresh paper until the oil will no longer drain from the

FIG. 22. MODE OF HANDLING THE COVER-GLASS IN MOUNTING TISSUES—FREEBORN.

section when the slide is held vertically. With a glass rod remove a little of the xylol balsam (*vide* formulæ) from the bottle, and allow a drop of this balsam to fall upon the section.

Pick up a clean cover-glass with the forceps, and place it on the drop of balsam. This operation is seen in Fig. 22. The point of the forceps may be placed beneath the cover-glass, the tip of the forefinger pressing lightly over it, and you will be enabled to carry the thin glass wherever desired.

As the cover settles down the air is pressed out, until finally the section appears imbedded in the varnish—the latter filling the space between the cover and the slide.

The object is "mounted." You have a *permanent specimen*. The slide must be kept flat, as the balsam is soft. After some weeks, the varnish around the edges of the cover will stiffen, and eventually become solid. Do not paint colored rings around the specimen. Nothing can present a neater appearance than the simple mount, as we have described it, after having been properly labeled. Labels seven-eighths of an inch square may be put on one or both ends, with the name of the object, date, method of staining, or whatever particulars you may prefer.

Specimens should be kept in trays or boxes in such manner that they may always lie flat.

CARE OF THE MICROSCOPE

The objectives constitute the most valuable part of the instrument. The lenses should never be touched with the fingers; indeed, the same rule applies to all optical surfaces. When the glasses become soiled they may be cleaned, but it should be done with great care. While the effect of a single cleaning would probably not be of the slightest appreciable injury to the glass, repeated wiping with any material, however soft, will destroy the perfect polish, and result in obstruction of light and consequent dimness in the field. Never use a chamois leather on an optical surface, as these skins contain gritty particles. Old, well worn linen and Japanese paper are by far the best materials for wiping glasses. If a lens be covered with dust, brush it off, breathe on the surface, and wipe gently with the linen or paper. Should you get clove oil on the front lens of the objective (as frequently happens when examining temporary mounts) wipe it dry, and then clean with the linen moistened with a drop of alcohol. Canada balsam can be very readily removed from any surface after having softened it with oil of cloves. The front lens of the objective, being the only one exposed, is the one usually soiled.

Particles of dirt on the objective, as I have said, cause a dimness in the field—the image is blurred. Dust on the lenses of the eye-piece, however, appears in the field. These lenses are readily cleaned by dusting and wiping with the linen, after having breathed on the surface. Never wipe a lens when dusting with a camel's-hair brush will answer the purpose.

The microscope should either be covered with a shade or cloth, or put away in its case when not in use. The delicate mechanism of the fine adjustment becomes worn and shaky if not kept free from dirt.

PART SECOND

STRUCTURAL ELEMENTS

PRELIMINARY STUDY

FORM OF OBJECTS

From a single and hasty view of bodies under the microscope, we are liable to form erroneous ideas of form. Either a sphere, disc, ellipsoid, ovoid, or cone may be so viewed as to present a circular outline. It therefore becomes important to view objects in more than a single position. This can easily be accomplished with isolated particles by suspension in a liquid. In this way the true shape of a blood-corpuscle, *e.g.*, may be determined.

Again, much information concerning the actual form of bodies may be gained by a proper adjustment of the fine focusing screw. You may remember that the depth of the field of view in the microscope is exceedingly slight. Speaking accurately, only a single plane can be seen with a single focal adjustment; but by gradually raising or lowering the tube of the microscope the different parts of a body may be focused and studied, and an accurate idea of form secured.

With a glass rod place a drop of milk, which has been previously diluted with three parts of water, on a slide, and put a cover-glass thereon, as in Fig. 23. Focus first with the low-power (L). A multitude of minute dots are observed. Then change to the high-power (H), and the dots will resolve into circular figures. Select one of the smaller particles, and, as you raise the focus, the center of the figure retains its brilliancy, while the edges become dark or blurred, showing convexity. Reverse the focus, and the center again retains its sharpness long after the edge has become blurred. The figure, then, is a spheroid. These bodies are fat-globules. Particles of free fat always assume the spheroidal form when suspended in a liquid.

Note the larger globules: they have become flattened by the pressure of the cover-glass.

Clean the slide, and make a second preparation from the diluted milk—first, however, shaking it violently in a bottle. Note the flattened air-bubbles among the oil-globules. Observe that these air-bubbles have no intrinsic color, while the fat-globules are

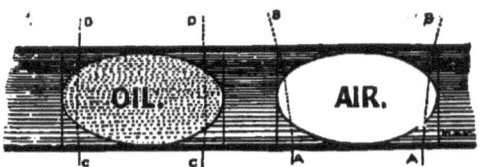

Fig. 23. Diagram showing the effect of Air-bubbles and Oil-globules in a mounted specimen upon the rays of light.

The lines A, B show the refraction of the rays (so as to produce a ring of color) by the action of two plano-concave water-lenses which are formed by the air-bubble.

The oil is seen to correct the refraction of C, D, thus giving but little color to the margin of this globule.

faintly yellow. Observe the change in the ring of prismatic color about the edge of the air-bubble, as the focus is altered. No such color will be seen in connection with the oil-globule.

The bubbles assume various figures from the pressure of the cover-glass.

MOVEMENT OF OBJECTS

Objects are frequently seen moving in the field of the microscope, the movement being magnified equally with their dimensions.

Thermal Currents.—When, with the previous specimen or any other fluid mount, the warm hand is brought close to one side of the stage, the globules in the field will be seen swimming more or less rapidly. These currents are due to the unequal heating of the liquid under observation. The direction of the current is in the reverse of its apparent motion.

Brownian Movement.—Place a fragment of dry carmine on a slide, add a drop of water, and with a needle stir until a paste is formed. Add another drop of water, and immediately put on the cover-glass. With H, note the most minute particles, and observe their peculiar, dancing motion. This occurs when almost any finely divided and generally insoluble solid is mixed with water. It ceases after a short time. The movement has been attributed to several causes.

Vital Movements.—Place a drop of decomposing urine on a slide, cover, and focus with H. The field contains innumerable minute spherules and rods (bacteria) which are in active motion, resembling somewhat the Brownian movement, although sufficiently distinctive after close observation.

After having rubbed the tongue for a moment against the inner surface of the cheek, put a drop of saliva on a slide, cover, and focus (H). Among the numerous thin, nucleated scales and debris, small granular spherules—the salivary corpuscles—will be found. Select one of the last, center, and focus (H) with extreme care. The minute granules within the cells are in active motion, resembling the Brownian movement, but with proper conditions the motion may continue for many hours.

EXTRANEOUS SUBSTANCES

Before we begin the study of animal tissues, we wish to have you become somewhat familiar with the appearance of certain objects

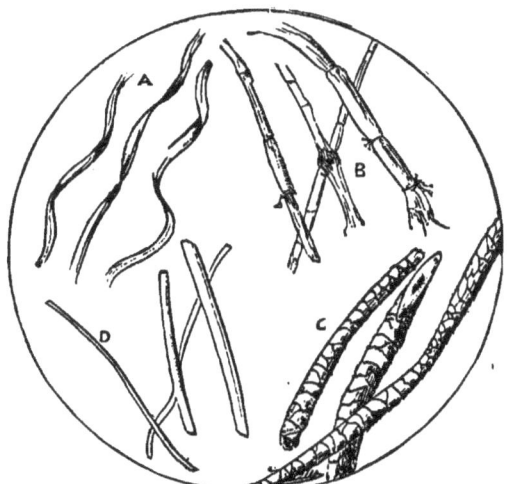

FIG. 24. EXTRANEOUS SUBSTANCES.

A. Cotton fibers, showing the characteristic twist.
B. Linen fibers, with transverse markings, indicating segments.
C. Wool. The irregular markings are produced by the overlapping of flattened cells. Wool may be distinguished from other hairs by the swellings which appear at irregular intervals in the course of the former.
D. Silk. Smooth and cylindrical.

which are frequently, through accident or carelessness, and often in spite of the utmost care, found mixed with our microscopical specimens. Among the more common objects floating in the air and gaining access to reagents, to subsequently appear in our mounted specimens, are the following:

Fibers.—Procure minute pieces of uncolored linen, cotton, wool, and silk. With a needle in either hand, tease out or separate a few fibers on slides, add a drop of water, and cover.*

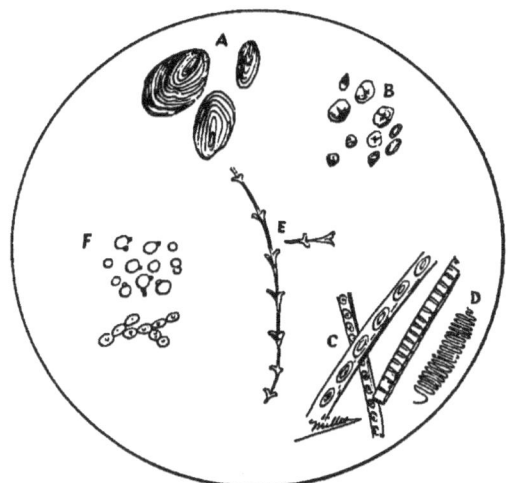

FIG. 25. EXTRANEOUS SUBSTANCES.

A. Granules of potato starch.
B. Corn starch.
C. Wood fibers. The circular dots are peculiar to the tissue of cone-bearing trees.
D. Spiral thread from a tea leaf.
E. Fragment of feather.
F. Cells of yeast and mould.

Starch.—Procure samples of wheat, corn, potato, and arrow-root starch, or scrape materials containing any one of these substances with a sharp knife. To a minute portion on the slide add a drop of water, cover, and examine with L and H.

Wood Shavings, Feathers, Minute Insects, Portions of Larger Insects, Pollen, etc., are easily mounted temporarily or permanently,

---

*These substances, as well as most of those which follow under the same heading, may be mounted permanently as follows: Put the dry material in clean turpentine for a' day or two, to remove the contained air. Transfer to the slide, tease, separate, or arrange the elements, after which wipe away the turpentine with strips of blotting-paper. Add a drop of balsam, and place the cover-glass thereon. The weight of the cover will be sufficient to press the object flat, if it be properly teased or separated.

as before noted. They are very commonly found in urine after it has been exposed to the air, and their recognition is very important.

Let me urge you to become familiar with the microscopical appearance of the commoner objects which surround us in everyday life. The most serious mistakes have resulted from ignorance of this subject. Vegetable fibers have been mistaken for nerves and urinary casts, starch granules for cells, vegetable spores for parasitic ova, etc.

STRUCTURAL ELEMENTS

Certain anatomical structures, of a more or less elementary nature, are united in the composition of organs. These structural elements will, with propriety, first claim notice from us.

CELLS

A typical cell is a microscopical sphere of protoplasm, constituted as follows (*vide* Fig. 26):

A. Limiting membrane.
B. Cell-body.
C. Nucleus.
D. Nucleolus.

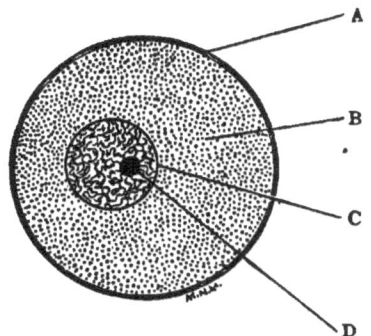

FIG. 26. ELEMENTS OF A TYPICAL CELL.

The *wall* consists of an apparently structureless membrane of extreme tenuity.

The *cell-body* may be either clear (jelly-like), granular, or fibrilliated. It contains an albuminous substance called *protoplasm*.

The *nucleus* is a minute spherical vesicle, with a limiting membrane inclosing a clear gelatinous material, traversed by a reticulum of fibrillæ.

The *nucleolus* consists of a spherical, highly refracting body lying inside of the nucleus, sometimes appearing to be a granular enlargement upon the fibrillæ of the nucleus.

Deviations from the type are most frequent, and vary greatly as to form, number of elements, and chemical composition.

Fig. 27. A Cell Nucleus, with Network and Nucleolus. Diagrammatic.

The typically perfect cell is rarely seen in human tissue on account of the length of time which commonly elapses between death and observation of the structure, the delicate fibrillæ of the nuclei usually appearing as a mass of granules.

CELL DISTRIBUTION

The complex mechanism of the body had its origin in a single cell. This preliminary structure, endowed with the power of proliferation, became two cells. Two having been produced, they became four; the four, eight; and thus progression advanced until they became countless. Some of these cells remained as such; others, altered in form and composition, gave birth to muscle, bone, etc. The study of these processes belongs to physiology.

The adult body is composed largely of cells of various forms. The different physiological processes, as secretion, absorption, respiration, etc., are effected through the intervention of these anatomical elements.

All free surfaces, within or without the body, *are covered with cells*. The entire skin, the outside of organs, as the lungs, liver, stomach, intestines, brain, etc.; all cavities, as the alimentary tract, heart, ventricles of the brain, blood-vessels and ducts, present a superficial layer of cells.

The cells are held together by an intercellular substance, which may be so abundant that the cells form a comparatively small part of the tissue. The intercellular material is to be regarded as having been made by the cells. In the case of the connective tissues, it has a more important function than the cells; while the amount of intercellular substance in epithelial structures is trifling.

CELL–DIVISION

The increase in the number of cells in the body may be accomplished in two ways:

a. *By direct division.*
b. *By indirect division.*

Direct division is a process in which the cell becomes constricted and a portion of it separated from the remainder. It is

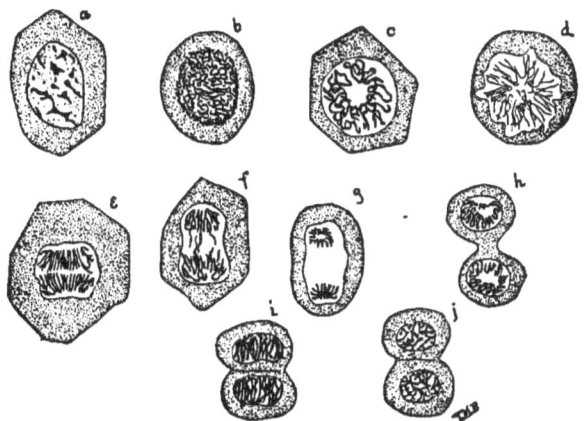

FIG. 28. INDIRECT CELL-DIVISION—AFTER FLEMMING.

now believed that in most instances the separation of the protoplasm is preceded by a series of changes in the nucleus. This mode of division is called indirect. The changes in the nucleus go by the name of karyokinesis, or mitosis.

KARYOKINESIS

The reticulum of fibers already mentioned as traversing the substance of the nucleus has an affinity for nuclear stains, and its substance is therefore called *chromatin*. During karyokinesis the chromatin of the nucleus undergoes a series of complicated changes, resulting in its division into two equal parts. This is followed by the division of the rest of the cell. During the process the chromatin presents figures of great variety and intricacy.

The effect of these changes is to separate the chromatin into masses called *chromosomes*. The number of chromosomes varies in different species, but is probably constant in the same species. Fig. 29 shows the main events in karyokinesis in a diagrammatic manner. It represents the process as it occurs in the starfish, where the process is less complicated than in some cases. Each

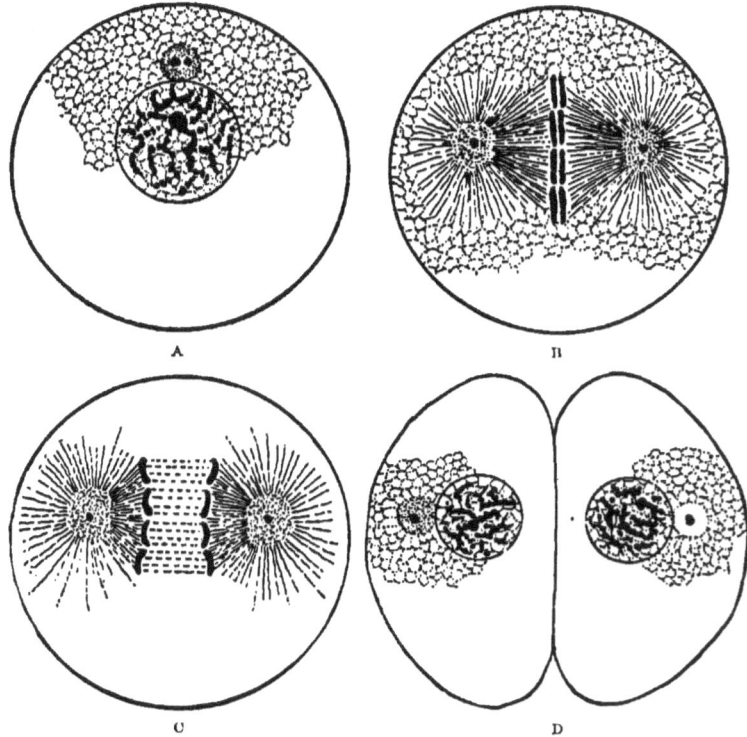

FIG. 29. KARYOKINESIS—AFTER WILSON.

chromosome becomes split lengthwise into two halves. One group of halves moves to one end or *pole* of the cell, the other group to the other pole. The two groups of chromosomes give rise to two new nuclei.

The separation of the groups of chromosomes is accomplished by delicate filaments, which radiate from the two poles to which the chromosomes are to travel. Some of these filaments meet at the *equator* of the cell to form what is called the *nuclear spindle*. The radiating filaments at the poles of the cells present figures which have been compared by Wilson to the arrangement of iron filings about the poles of a horseshoe magnet. The circle of radiating filaments at each pole is called the *attraction-sphere*, the center of which is the *centrosome*. It is possible that the substance of the filaments serves to draw the chromosomes apart. Apparently the centrosome is to be regarded as a permanent organ of the cell. Its division must precede the division of the nucleus.

The time required for cell division in man is about half an hour (Stöhr).

<small>Most of the stages in karyokinesis may be demonstrated in the epithelial cells in the tail of a young newt or salamander tadpole. Fix in Flemming's osmic acid or chromic-acetic solution for twenty-four hours; wash; stain with safranin or hæmatoxylin; and, after dehydrating and clearing, mount in balsam. Examine with oil-immersion lens, if possible.</small>

CLASSIFICATION OF TISSUES

Histology, or microscopical anatomy, treats of the minute structure of the tissues and organs of the body.

A tissue is a collection of similar cells and intercellular substances. The principal tissues are epithelial, connective, muscular, and nervous. Blood is sometimes considered as a tissue. Most organs are made of several different tissues. The following table shows the varieties of tissues:

Epithelial Tissue { Squamous. Columnar.

Endothelium.
Blood.

Connective Tissue { Mucoid. White Fibrous. Yellow Elastic. Adipose. Retiform (of lymphoid tissue). Cartilaginous. Osseous. Dentine.

Muscular Tissue { Unstriated. Striated. Cardiac.

Nervous Tissue.

EMBRYONIC DERIVATION OF TISSUES

In correlating the work of histology with that of embryology, the student will find the following table serviceable. The table indicates from which layer of the blastoderm each of the principal tissues is derived:

Ectoderm or Epiblast
- The epithelium of the skin and all its appendages and glands.
- The epithelium covering the front of the eye, the crystalline lens, and the retina.
- The epithelium of the external auditory canal and of the membraneous labyrinth.
- The epithelium of the nasal cavity and its diverticula.
- The epithelium of the mouth and its glands, the salivary glands, and the enamel of the teeth.
- The epithelium of the anal end of the alimentary canal.
- The tissues of the nervous system, the lining of the central canal of the spinal cord and of the cerebral ventricles, the pituitary and pineal bodies.

Mesoderm or Mesoblast
- All connective tissues.
- Muscular tissue.
- The blood-and lymph-vessels and their endothelium, and the endothelium of the serous membranes.
- Blood-and lymph-corpuscles.
- The spleen and lymph-nodes.
- The kidney and ureter.
- The ovary and testicle and their ducts, except the external ends of the ducts.

Entoderm or Hypoblast
- The epithelium of the alimentary canal (except at the upper and lower extremities), and all the glands opening into it, including the liver, pancreas, thyroid, and thymus.
- The epithelium of the respiratory tract which originates as a diverticulum from the alimentary canal.
- The epithelium of the Eustachian tube and tympanum.
- The epithelium of the urinary bladder and urethra.

EPITHELIUM

Epithelium is the tissue covering the surfaces of the body that communicate with the external world, and lining the glands.* It is made of cells, held together by only a small amount of intercellular substance.

*Note the exceptions in the case of the peritoneum, which is lined by *endothelium*, and which opens externally by means of the Fallopian tube; the thymus and thyroid glands, and the cavity of the central nervous system, which contain or are lined by epithelium, but do not open externally.

Where a layer of epithelial cells comes in contact with the underlying connective tissue, the deepest epithelial cells often rest upon a thin *basement membrane*, which is modified connective tissue, either structureless or made of flattened cells.

The succeeding pages will show that epithelial cells may be — squamous or columnar, ciliated or otherwise, simple or stratified. These varieties of epithelium are distributed as follows, giving only their most important locations:

Simple Squamous Epithelium
- The alveoli of the lungs.
- The capsule of the Malpighian body and the descending limb of the loop of Henle in the kidney.

Stratified Squamous Epithelium
- The covering of the skin, of the eye, mouth and tongue, pharynx, œsophagus, epiglottis, and of the upper part of the larynx.
- The lining of the urinary tract from the pelvis of the kidney down (except part of the male urethra), most of it being of the special variety of stratified squamous epithelium known as *transitional*.
- The lining of the vagina.

Simple Columnar Epithelium
- The alimentary canal from the beginning of the stomach to the lower part of the rectum; the ducts of its glands and of many other glands; the seminal vesicles, ejaculatory ducts, and part of the male urethra; the surface of the ovary.
- It is ciliated in the uterus and Fallopian tube, the central canal of the spinal cord, and the cerebral ventricles.

Stratified Columnar Epithelium
- It is ciliated in the most important situations: The trachea and bronchial tubes, the Eustachian tube, the upper part of the pharynx, the lower part of the nasal cavity, the vas deferens.

SQUAMOUS, STRATIFIED, AND TRANSITIONAL EPITHELIUM

The simplest method of tissue production by means of flat cells is that of superposition, constituting *squamous epithelium*. Cells are placed one over the other, generally without great regularity. If regular, and in several layers, the structure is called *stratified* epithelium; if only in a few layers, it is termed *transitional* epithelium. The superficial layer of the skin affords an example of squamous, stratified epithelium. The bladder and pelvis of the kidney are lined with transitional epithelium.

The thin, flat scales from the mouth may be demonstrated by scraping a drop of saliva from the tongue with the handle of a scalpel, transferring it to the slide, and applying the cover. The size of the drop of saliva should be carefully adjusted so as to fill the space between the cover-glass and slide. Too little will cause the cover to adhere so tightly to the slide as to press the cells out of form; too much, and the saliva flows over the cover and soils the objective. With a glass rod place a drop of the dilute eosin solution on the slide, and with a needle lead it to the edge of the saliva. The dye will pass under the cover slowly, and whatever anatomical elements there may be present will be gradually stained. Observe that the nuclei of the flat scales first take the dye and appear of a deep pink; while the other portions are either colorless or very lightly stained.

Find a typical field and sketch it with a pencil, afterward tinting with dilute eosin.

An admirable view of the surface of a squamous epithelium may be had by using the superficial layer of the skin of a frog, which is

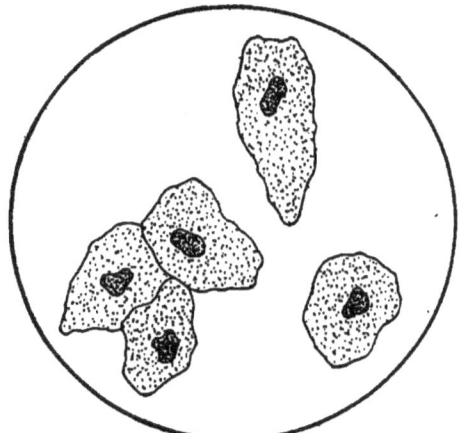

Fig. 30. Squamous Epithelial Cells from the Mouth.

often shed when the animal is kept in confinement. It may be stained with hæmatoxylin; and small pieces, after alcohol and clearing, may be mounted in balsam.

PAVEMENT EPITHELIUM

When thin, flat cells are disposed in a single layer, like tiles, the epithelium is termed simple squamous, pavement, or tessellated. These cells are often quite regularly polygonal (although this obtains more frequently with tissue from the lower animals), and they are always connected by their edges by means of an albuminous cement.

Such a simple squamous epithelium is found in only a few locations. The most important are the alveoli of the lungs, the

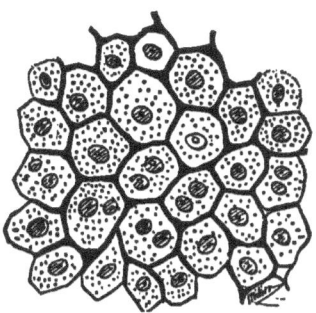

FIG. 31. PAVEMENT EPITHELIUM. DIAGRAMMATIC.

capsule of the Malpighian body of the kidney, and the descending limb of the loop of Henle in the kidney. Seen from the surface, a stratified or columnar epithelium appears like pavement epithelium.

COLUMNAR EPITHELIUM

Columnar cells are found, generally, throughout the alimentary and respiratory tracts, except near their external openings. They also line the cerebral ventricles, most of the uriniferous tubules, Fallopian tubes, the uterus, etc. This epithelium is quickly destroyed after death, and is difficult of perfect demonstration, except in an animal recently killed.

Procure from the abattoir a portion of the small intestine and bronchus of a pig, and with the curved scissors snip out small pieces from the mucous surface of each. Macerate in one-sixth per cent. of chromic acid for twenty-four hours.

Place a piece of the gut on a slide, and, after having added a drop of the acid solution, scrape off the mucous surface with a knife and remove the remainder of the gut. Add a cover-glass, and focus (H). You will find cells in various conditions, from isolated examples to small groups like Fig. 32.

Observe that the attached ends of the cells are often small and pointed, and that spheroidal and ovoidal cells are frequently wedged in between them. Note the free border: it consists of striæ, and

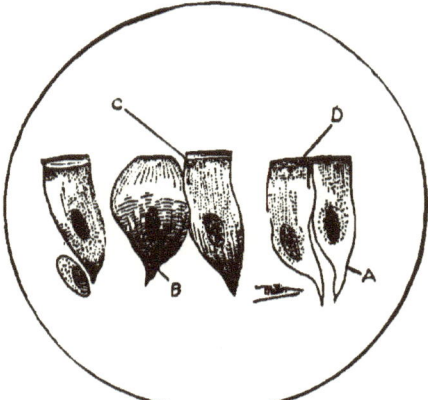

FIG. 32. COLUMNAR CELLS FROM SMALL INTESTINE OF RABBIT.
A. Tapering attached extremity.
B. A swollen goblet cell.
C. Finely striated free border.
D. Transparent line of union between the striated portion and the body of the cell ($\times$ 400).

is separated from the body of the cell by a translucent line. This appearance is also that of the epithelium in the human intestine.

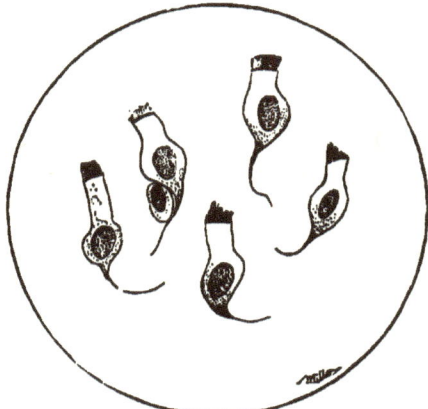

FIG. 33. CILIATED COLUMNAR CELLS FROM BRONCHUS OF PIG ($\times$ 400).

Ciliated Columnar Epithelium

Prepare, by scraping, a slide from the mucous surface of the pig's bronchus (which has been macerating in the chromic acid)

Observe the cilia on the free border of the cells. Interspersed between ciliated cells, much-enlarged individuals may be found— the so-called beaker, goblet, or mucous cells.

The motion of the cilia may be demonstrated as follows:

Carefully open an oyster so as to preserve the fluid. On examination you will notice the gills, shown in Fig. 34, commonly

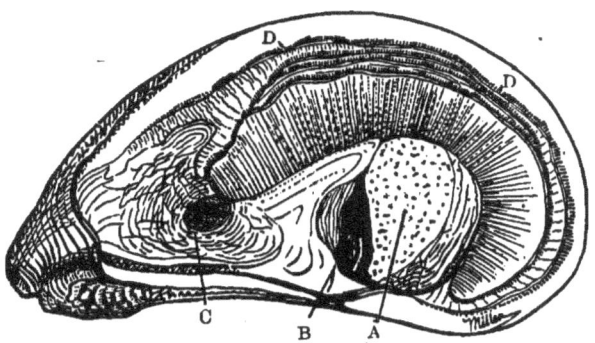

FIG. 34. OYSTER, OPENED TO SHOW METHOD OF PROCURING LIVING CILIATED CELLS.

A. The divided muscle. This must be sectioned before the shell can be opened
B. The heart.
C. Liver.
D, D. The so-called "beard." These laminæ are covered with cells provided with cilia; and a fragment of the free border of one of the leaflets may be snipped off with the scissors and examined as described in the text.

called the beard. With the scissors snip off a fragment of the free border of this beard, add a drop of the liquid from the oyster, and tease with a pair of needles. Apply the cover, and focus (H).

At first the individual cilia cannot be demonstrated on account of their rapid vibration. After a few moments, however, the action becomes less energetic, and the hair-like appendages of the cells are to be plainly seen.

GLANDULAR EPITHELIUM

Scrape the cut surface of a piece of liver; place the scrapings on a slide; add a drop of normal salt solution (*vide* formulæ); mix with a needle, and put on the cover-glass.

With H observe, among the numerous blood-corpuscles, fat-globules, etc., the *polyhedral liver-cells*, about twice or three times the diameter of a white blood-corpuscle (Fig. 35). Notice the large spherical nuclei, with nucleoli. Note, also, the yellow

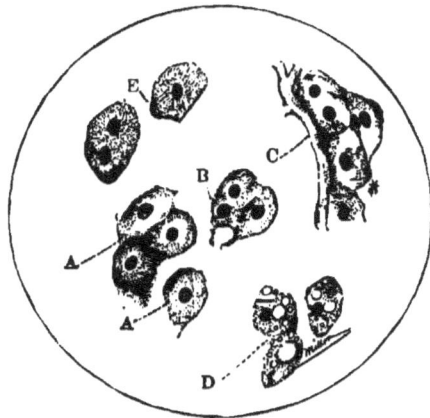

Fig. 35. Glandular Epithelium.

A, A. Polyhedral cells from human liver.
B. Double nuclei.
C. Cells from same showing connection with a capillary.
D. Same cells infiltrated with globules of fat.
E. Cells from liver of pig showing intracellular network (× 400).

pigment-granules and the fat-globules in the body of the cells. Masses of these cells resemble somewhat pavement epithelium; they are not flat, but polyhedral.

ENDOTHELIUM

Endothelium resembles simple squamous epithelium in appearance, and it is called epithelium by some histologists.

Endothelium forms the superficial covering of the pleural, pericardial, and peritoneal cavities, the tunica vaginalis, and the joints. It covers the membranes of the brain and spinal cord, the inner surfaces of the heart, and the blood- and lymphatic vessels. Therefore endothelium lines the cavities that do *not* communicate with the external world.*

It consists of a single layer of thin, flat cells, held together by a small amount of intercellular substance. It is demonstrated best on fresh tissues, hence the human subject is seldom available.

The mesentery of the frog affords a good example of endothelial

---

*Remember that the peritoneum may be said to have an opening by way of the Fallopian tube, and that the central canal of the spinal cord and brain is lined by epithelium but has no opening that connects it with the external world.

structure, and differs but little from the arrangement on human serous surfaces.

Kill a large frog by decapitation, and open the abdomen freely by an incision along the median line. Pull out the intestines by grasping the stomach with the forceps. This will expose the small intestine, which you will remove, together with the attached mesentery, by means of quick snips of the scissors. Work as rapidly as possible and avoid soiling the tissue with blood. Throw the gut into a saltcellar filled with silver solution (*vide* formulæ), where it must remain for ten minutes covered from the light. Lift the tissue from the solution by means of a strip of glass (or a platinum wire), and throw into a saucer of clean (preferably distilled) water, changing the latter repeatedly for some minutes.

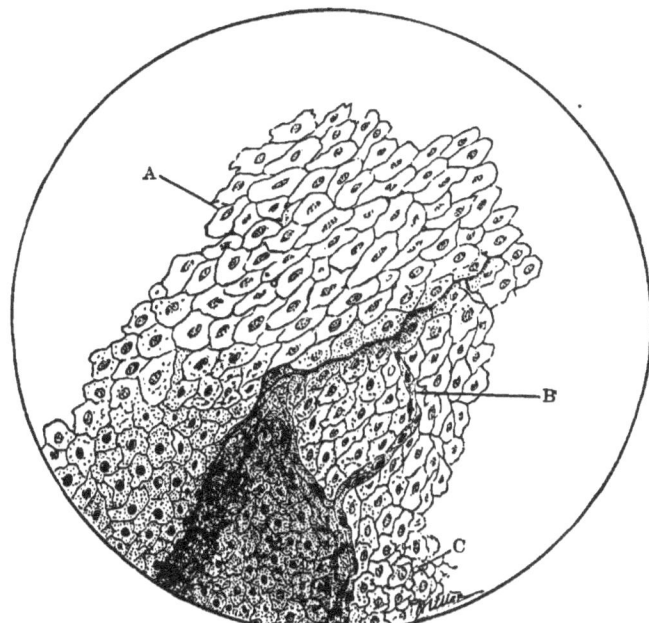

FIG. 36. ENDOTHELIUM FROM FROG'S MESENTERY. SILVER STAINING.

A. Area showing the outlining of the cells by the silver stained cement-substance. The nuclei have been brought out by the carmine. Minute stomata may be seen between certain cells.

B. A blood-capillary terminating below in an arteriole. The silver has outlined the endothelial cells of the vessels.

C. An area showing both layers of the cells. The deeper cells are faintly outlined, being out of focus. The silver has been deposited over the lower portion of the specimen, nearly obscuring the cement lines ($\times$ 250).

After washing, and while yet in the water, expose to sunlight (perhaps fifteen minutes) until a brown tint is acquired, which indicates the proper staining.

The mesentery has been left connected with the intestine, so that the former might not curl. The preparation having been dehydrated with alcohol, and having reached the oil of cloves, proceed with a pair of scissors to snip off a small, flat piece of mesentery. Remove it to a slide and mount in balsam. The omentum of a rabbit, cat, or dog may be prepared in a similar manner.

The outlines of the endothelial cells appear as dark lines where the silver has stained the cement-substance. These lines are often tortuous and serrated. The nuclei are not seen, unless they have been separately stained with carmine or some other nuclear dye.

Openings called *stomata* occur on the serous surfaces, which connect the cavities with underlying lymphatic vessels. The stomata occur at the points where several endothelial cells meet. The opening is sometimes surrounded by several small, granular cells, called "guard cells." Changes in the size of these cells modify the size of the stomata.

CONNECTIVE TISSUES

Certain elementary structures of similar origin and mode of development, and serving alike to unite the various parts of the body, have been termed connective tissues. Custom has restricted the term, in its everyday employment, so as to apply to white fibrous tissue, or, at least, to tissue which always resembles this more closely than any other.

WHITE FIBROUS TISSUE

This, the connective tissue *par excellence*, is composed of exceedingly fine fibrillæ (only $0.6\,\mu$ in diameter), which are aggregated in irregularly sized and variously disposed bundles. It forms long and exceedingly strong tendons connecting muscle and bone; its fibers interlace, forming the delicate network of areolar tissue; it forms thin sheets of protecting and connecting aponeuroses;. or, supporting vessels, it permeates organs and sustains the parenchyma of glands.

The fibers are held together by means of a transparent cement, which may be softened or dissolved in acetic acid. They may exist, as in dense tendons, without admixture.

Cells are found between the bundles of fibers, known as connective tissue corpuscles. The older and more dense the structure, the less frequent are these cells; while in young connective tissue, stained, the nuclei of the corpuscles constitute a prominent feature of the specimen under the microscope.

Having obtained a piece of tendon from a recently killed bullock, tease a fragment on a slide in a few drops of water. Select a

FIG. 37. CONNECTIVE TISSUE.
A. Teased fibers.
C. Fibrillæ.

portion which splits easily and separate the fibrils as much as possible. Cover, and examine (H).

Fine, wavy fibers are seen composing the fasciculi. If the dissection has been sufficiently minute, you may succeed in demonstrating ultimate fibrillæ. These are best made out, as at C in Fig. 37, where the parts of a bundle have been separated for some distance, leaving the finer elements stretching across the interval.

YELLOW ELASTIC TISSUE

This tissue consists of coarse, shining fibers (averaging about 8 μ in diameter) which frequently branch and anastomose. They are

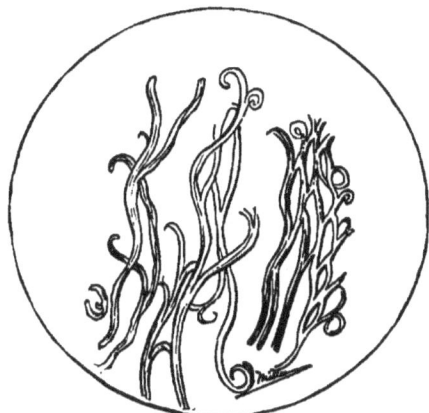

FIG. 38. TEASED YELLOW ELASTIC TISSUE FROM THE LIGAMENTUM NUCHÆ. (× 250.)

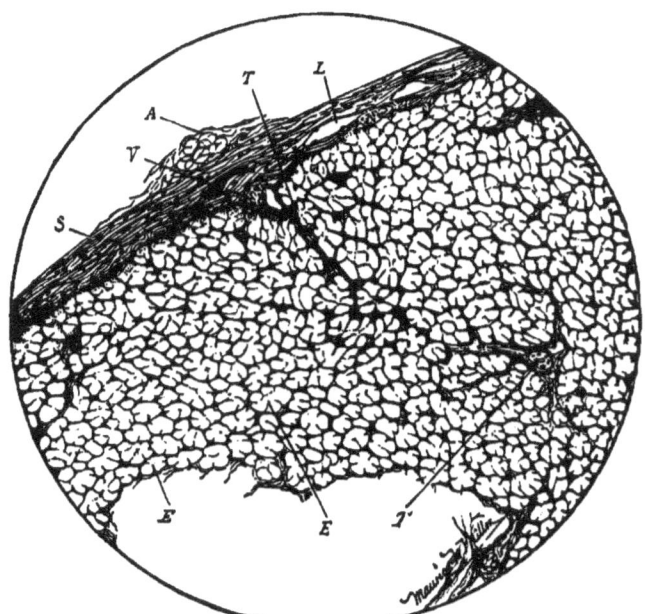

FIG. 39. TRANSVERSE SECTION OF PART OF THE LIGAMENTUM NUCHÆ.

S. Sheath of the ligament, sending prolongations within—as at T, T—dividing the structure into irregular bundles or fasciculi.

L. Lymph-spaces in the connective tissue.

A. Adipose tissue in the sheath.

V. Blood-vessels in transverse section.

E, E. Primitive fasciculi of yellow elastic tissue fibers.

highly elastic. Under the microscope the fibers are colorless, but when aggregated, as in a ligament, the mass is yellow.

Procure a small piece of the *ligamentum nuchæ* of the ox, and tease it on the slide after it has been macerated in acetic acid for a few moments. The acid softens the fibrous connective tissue and facilitates the teasing process.

The individual fibers having been isolated, they appear as in Fig. 38. When broken, they curl upon themselves, like threads of India rubber.

This tissue is variously disposed throughout the body where great strength with elasticity becomes necessary. The large arteries are abundantly supplied with elastic fiber, arranged in plates, in alternation with muscle. As a network, it is mixed with connective tissue in the skin, and in membranes generally. It contributes elasticity to cartilage where the fibers form an intricate network.

Ligaments are composed largely of yellow elastic tissue. Fig. 39 is drawn from a portion of a stained transverse section of part of the *ligamenta subflava*.

A strong *sheath* of fibrous tissue is thrown around the whole ligament, a portion of which is seen at S. This sheath sends prolongations, T, T, into the structure, dividing it into irregular bundles, which support nutrient vessels. The elastic fibers seen in transverse section, as at E, E, are observed strongly bound together with fibrous tissue, which penetrates the smaller fasciculi, dividing them into the *ultimate fibrillæ*.

ADIPOSE TISSUE

Adipose or fat tissue is a modification of and development from ordinary connective tissue.

It originates in certain contiguous connective tissue corpuscles becoming filled with minute fat-globules. These ultimately coalesce and form single large globules, which bulge out the cell-bodies until they become spheroids; the nuclei at the same time are displaced to the periphery. An aggregation of such cells forms a lobule of adipose tissue. The cells are often so closely packed as to assume a polyhedral form. From malnutrition, this fat may be absorbed, ordinary connective tissue remaining.

You will bear in mind the fact that whenever fat exists in a condition of minute subdivision, the particles always assume the

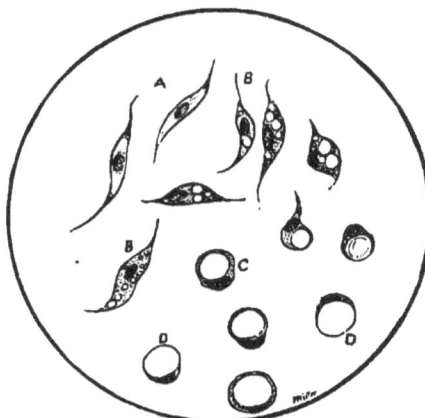

Fig. 40. Connective tissue Cells containing Fat—indicating the mode of formation of Adipose Tissue ($\times 400$).

A. Ordinary elongate connective tissue cells.
B. Same containing minute globules of fat.
C. Coalescence of the fat-globules and displacement of the nucleus.
D. Still greater increase of the fat.

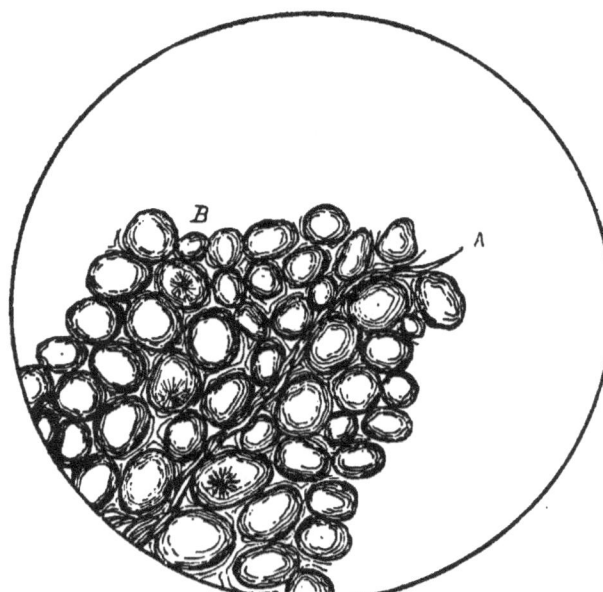

Fig. 41. Adipose Tissue from Teased Human Omentum. Stained with Hæmatoxylin ($\times 400$).

A. Connective tissue framework.
B. Just below the letter is a cell containing crystallized fat

globular form; and that while adipose tissue contains fat, fat alone is not adipose tissue.

CARTILAGE

Cartilage consists of a dense basis substance, in which cells or corpuscles are imbedded. It presents three forms:

HYALINE CARTILAGE

The matrix of hyaline cartilage is translucent, dense, and apparently structureless. Minute channels in certain instances and delicate fibrillæ in others have been demonstrated. The surface of the cartilage is surrounded by a fibrous envelope, called the *perichondrium*.

The basis material contains excavations, generally spherical, called *lacunæ*. They appear to be lined with a membrane, and

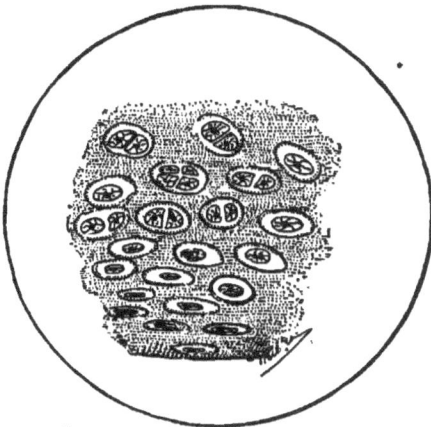

Fig. 42. SECTION OF HYALINE CARTILAGE FROM A HUMAN BRONCHUS.

The ground-substance is apparently structureless, and it contains the lacunæ or excavations in which one, two, three, or more cartilage cells appear. These cells show a well marked intracellular network ($\times$ 400).

contain one, two, three, and perhaps as many as eight cells— the *cartilage-corpuscles*. The matrix around the lacunæ is called the *capsule*. It differs from the rest of the matrix and creates the appearance of a membrane surrounding the lacunæ.

Hyaline cartilage is found covering joints generally, where it is termed articular cartilage. It is also found in the trachea, the bronchi, the septum narium, etc.

Fig. 42 shows a section of hyaline cartilage from one of the rings of a large bronchus.

FIBRO-CARTILAGE

Fibrous connective tissue, predominating largely in the basis substance, produces a structure of great strength—fibro-cartilage. The intervertebral disks afford an example of this variety, from a

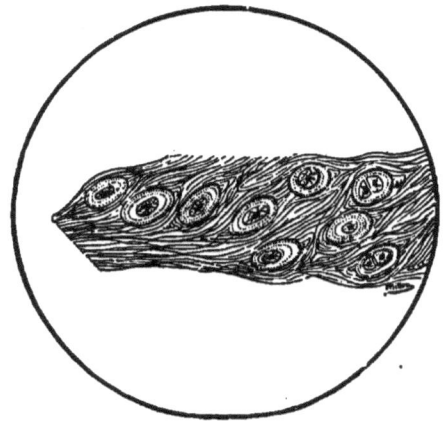

FIG. 43. FIBRO-CARTILAGE FROM AN INTERVERTEBRAL PLATE OR DISK.
The ground-substance, unlike that of the hyaline variety, consists of dense fibrous tissue with little calcareous matter (×400).

section of which Fig. 43 has been drawn. The fibrous tissue is a very prominent feature of the ground substance.

ELASTIC OR RETICULAR CARTILAGE

As the name implies, yellow elastic tissue is an important element of the ground-substance of elastic cartilage. It presents the form of a reticulum, as shown in Fig. 44. It is not extensively distributed in the human being, although the cartilages of the external ear, Eustachian tube, epiglottis, etc., are of this variety.

Cartilage should be hardened by the chromic acid and alcohol process. The sections from which the illustrations have been drawn were cut without the microtome. They should be cut

CARTILAGE—BONE 69

Fig. 44. Elastic Cartilage from Ear of Bullock.
The ground-substance consists largely of a network of coarse, yellow elastic tissue (×400).

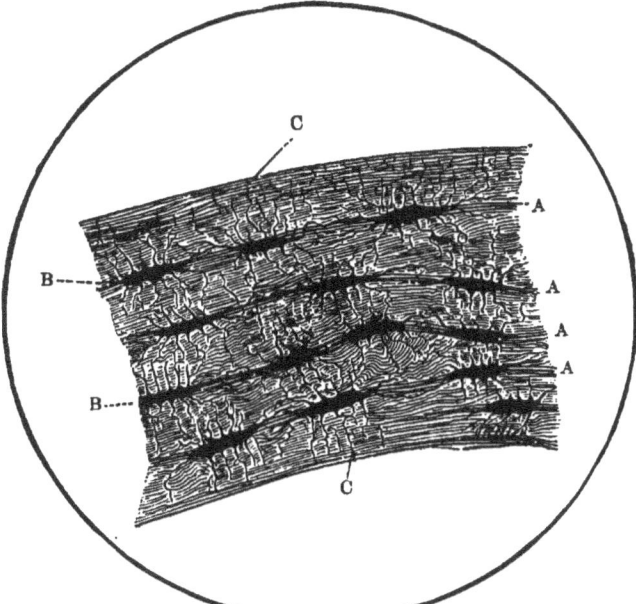

Fig. 45. Portion of a Transverse Section from a Dried Femur, Showing Part of the Wall of a Haversian System. (See pages 70 and 71.)

A, A. Bony lamellæ.
B, B. Lacunæ.
C, C. Canaliculi (×400).

extremely thin, but not necessarily large. We frequently succeed in getting good fields from the *thin edges* of sections which may be elsewhere too thick. Stain with hæmatoxylin and eosin. The differentiation will be excellent. The delicate nutritive channels in the matrix connecting the lacunæ may be demonstrated in the cartilage of the sternum of the newt; the xiphoid appendix is sufficiently thin without sectioning.

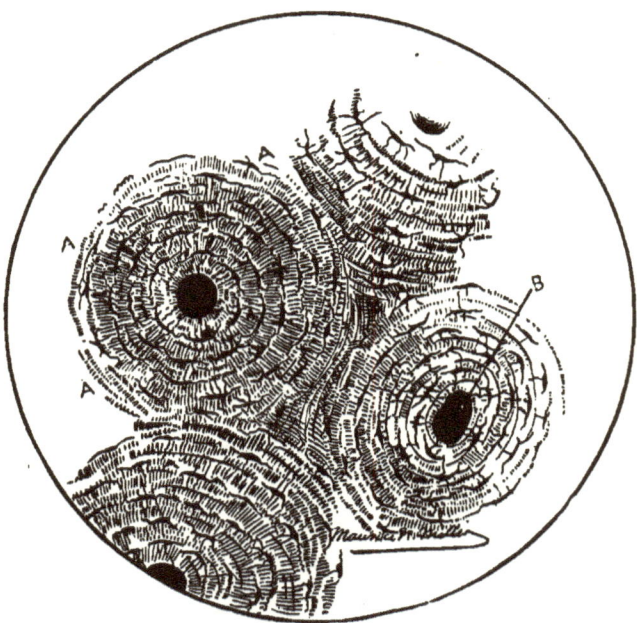

FIG. 46. TRANSVERSE SECTION OF PORTION OF A DRIED LONG BONE, SHOWING THE HAVERSIAN SYSTEMS.

A, A, A. A Haversian system.
B. Haversian canal.
The lacunæ, canaliculi, and Haversian canals all appear black in the section, as they are filled with air and the bony fragments resulting from grinding of the section (×60).

BONE

Bone consists of an osseous, lamellated matrix, in which occur irregularly-shaped cavities—*lacunæ*. The latter are connected by means of exceedingly fine channels—*canaliculi*. The lacunæ contain the *bone-corpuscles*, the bodies of which are projected into the canaliculi.

In compact bone, the blood-vessels run in a line parallel with the long axis of the bone, in branching inosculating channels (averaging about 50 μ in diameter)—the *Haversian canals*. The lamellæ of osseous tissue are arranged concentrically around these

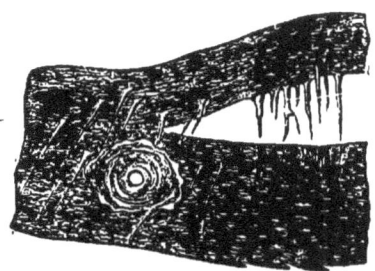

FIG. 47. SECTION OF BONE SHOWING SHARPEY'S FIBERS PULLED OUT OF POSITION. AFTER H. MÜLLER.

canals. A single Haversian canal, and the lamellæ surrounding and belonging to it, constitute a *Haversian system*.

The lamellæ beneath the periosteum are not arranged as above, but parallel with the surface of the bone. These plates are perforated at a right angle and obliquely by blood-vessels from the

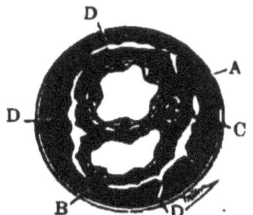

FIG. 48. DIAGRAM OF A HAVERSIAN CANAL.
A. Artery.
B. Vein.
C. Nerve.
D, D, D. Lymph-channels.

periosteum, as they pass on their way to the Haversian canals. These lamellæ are also perforated by partly calcified connective tissue—the *perforating fibers of Sharpey*. (Fig. 47.)

A Haversian canal contains (Fig. 48) an artery, a venule, lymph-channels, and a nerve-filament. The whole is supported by connective tissue cells with delicate processes. The walls of the lymph-spaces are prolonged into the canaliculi, and thus placed in connection with the elements of the surrounding lacunæ.

Each lacuna contains a bone-corpuscle, the protoplasmic body of which sends prolongations into the contiguous canaliculi. In the adult bone the cell is shrunken, and the processes just mentioned are not readily demonstrable.

The canaliculi of any Haversian system communicate with one another, but not with those of different systems.

The concentrically placed lamellæ around the Haversian canals are called *Haversian lamellæ*. The angular area formed where several Haversian systems join is filled out by *interstitial lamellæ*,

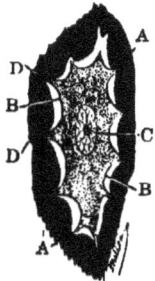

FIG. 49. DIAGRAM OF A BONE LACUNA.

A, A. Ground-substance of the bone.
B, B. Limiting membrane of the bone-corpuscle within the lacuna.
C. Nucleus and nucleolus of the corpuscle.
D, D. Projections of the cell-body into the canaliculi.

while those lamellæ mentioned above as occurring just below the periosteum are the *circumferential* or *fundamental lamellæ*.

The arrangement of Haversian canals and their concentric lamellæ is confined to *compact* bone. Compact bone is formed on the surface of all bones, and makes up the bulk of the shaft in long bones. The other variety of bone, called *cancellous* or *spongy*, occurs extensively at the ends of long bones, and in short and flat bones. It consists of lamellæ containing lacunæ and canaliculi, lying between good sized cavities, which are occupied by blood-vessels and marrow, and which correspond to Haversian canals. Bone is to be regarded as dense connective tissue, arranged in lamellæ, of which the ground-substance is impregnated with salts of calcium, chiefly the phosphate, to which its hardness is due.

At the same time, bone retains the elasticity and strength of connective tissue, which distinguish it from a structure that has merely been calcified, as may happen in certain disease processes.

PERIOSTEUM

The surface of bone is covered by an envelope called *periosteum*, which has two layers—a dense, outer, fibrous layer and a looser, inner, vascular layer. The inner layer contains the cells, *osteoblasts*, which form *osseous tissue*, hence this layer is called the *osteogenetic* layer of the periosteum. In operating on bone, surgeons guard the periosteum very carefully, on account of its blood-vessels and its osteogenetic elements.

MARROW

The spaces of bones are everywhere filled with marrow. The largest of these spaces are the medullary cavities of long bones. In the medullary cavities the marrow is yellow, owing to the deposit of fat in the cells. The smaller spaces of bone are filled with red marrow. Red marrow contains *marrow-cells*, which are similar to connective tissue cells, supported in a very vascular connective tissue framework. The marrow cells are identical with the osteoblasts of the periosteum and with bone-corpuscles. Immense cells, *giant cells*, containing many nuclei, also occur in red marrow. On account of their function of absorbing superfluous bone, they are called *osteoclasts* (Fig. 50). Red marrow also contains nucleated cells, tinged with hæmoglobin, that are connected with the formation of red blood-corpuscles, which probably takes place extensively in the red marrow.

DEVELOPMENT OF BONE

The majority of the bones of the body are first formed in the embryo as hyaline cartilage, which is subsequently replaced by true bone—*endochondral ossification*.

The bones of the face and of the vault of the cranium and a portion of the lower jaw are the principal exceptions—*intramembranous ossification*. In the latter case, the basis of the forming bone is an embryonic fibrous tissue, and this form of ossification differs from the endochondral in not being carried on in a cartilaginous basis, which is, however, a temporary structure.

(*a*) In the case of endochondral bone, the commencement of ossification is indicated by the enlargement of the cartilage-cells and their arrangement into vertical rows. This takes place at a point called the *center of ossification*. (*b*) The matrix between them becomes calcified. (*c*) From the surface of the bone, which is covered by a membrane corresponding to the periosteum, processes extend into the cartilage. (*d*) The cartilage is absorbed to make room for these processes through the agency of the osteoclasts (see Fig. 50). (*e*) The absorption proceeds until it includes the region of calcified cartilage. (*f*) The osteoblasts, which have accompanied the periosteal ingrowth, arrange themselves on the surface of the spaces resulting from absorption of the cartilage, and form layers of bone. (*g*) At the same time the osteoblasts of the osteogenetic layer of the periosteum form layers of bone beneath the periosteum—*periosteal ossification*.

The first bone formed is soft and spongy. It will be observed that as ossification proceeds the whole of the cartilage will be absorbed, except that at the epiphyses. In the course of the growth of the individual, the network and spaces of the original spongy bone undergo considerable rearrangement. Haver-

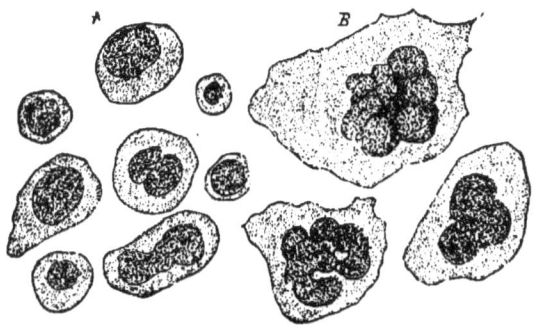

FIG. 50. CELLS FROM RED MARROW OF RABBIT (PRUDDEN).
A. Marrow cells proper.
B. Giant cells.

sian systems result from deposition of successive layers of lamellæ around the spaces of spongy bone, from without inward, leaving a channel at the center, which is the Haversian canal. The medullary cavity is produced by the absorption of the central part of the bone, while new layers continue to form under the periosteum.

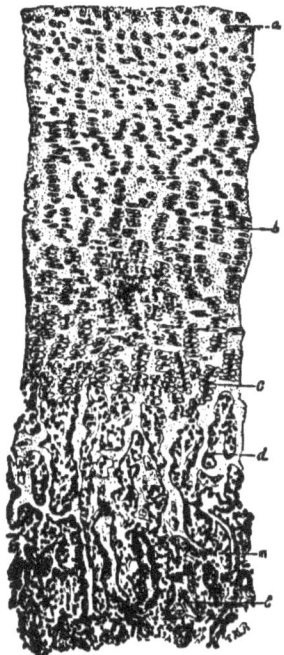

Fig. 51. Developing Bone, from the Pig (Prudden).

Fig. 46 has been drawn from a section of dry bone which has been sawn as thin as possible, and afterward rubbed down on a hone with water. It is a tedious process, and shows little but the osseous matrix. Bone should be decalcified for microscopical work, and it may then be readily cut in thin sections with a razor. The process is as follows:

To 100 c.c. of the dilute chromic acid solution add 3 c.c. of C. P. nitric acid. The bone, previously divided into slices not over one-half centimeter in thickness, is then placed in the fluid, and should be completely decalcified in a week or ten days. Examine the pieces after twenty-four hours by puncturing with a needle. Should the action proceed too slowly, add a few drops more of the nitric acid from time to time. The bone eventually takes on a green color. After complete decalcification, wash the pieces for twenty-four hours in clean water, and preserve them, until required, in 80 per cent. alcohol. Small pieces of young bone may be decalcified in a saturated aqueous solution of picric acid. The process is slow, but it leaves the tissue in excellent condition.

Sections cut in the usual way may be stained with carmine and picric acid, and examined in a drop of glycerin. They should not, after the staining, be placed in the oil of cloves, as they would curl and become hard. Transfer them to equal parts of glycerin and water, from which they are to be carried to the slide. Add a drop more of the dilute glycerin if necessary and put on the cover-glass, carefully avoiding air-bubbles. If you desire to make a permanent mounting, the edge of the cover must be cemented to the slide.

Thoroughly wipe the slide around the cover with moistened paper, *until every trace of glycerin is removed*. Then with a sable brush, paint a ring of zinc cement (*vide* formulæ) around the slide just touching the edge of the cover-glass. Repeat the cementing in twenty-four hours. A turn-table will be a useful aid in this work.

SPECIAL CONNECTIVE TISSUES

Connective Tissue of the Lymphatic System.—The matrix of lymphoid or adenoid tissue consists of a network of fibers and cells, which support the lymph-corpuscles. It is distributed extensively in organs, and where it appears in stained sections, the lymphoid cells are so numerous as to obscure the reticulum almost entirely. The structure will be minutely described in connection with the lymphatic system.

Embryonic Connective Tissue presents a homogenous, mucoid matrix containing branched cells. It is not found normally in the adult. The jelly of Wharton of the umbilical cord is *mucoid tissue*.

MUSCULAR TISSUE

This tissue is found in three varieties: 1. Non-striated, smooth or involuntary. 2. Striated, skeletal, or voluntary. 3. Cardiac.

NON-STRIATED MUSCLE

The histological element of non-striated muscle is a spindle-shaped cell from 45–225 μ long and 4–7 μ broad. The cell body presents longitudinal striæ, and contains an ovoid nucleus. The nucleus contains a reticulum which is probably in connection with the fibrillæ, which produce the longitudinal striation of the body. The cells are not infrequently bifid at one or both extremities. A transparent cement substance serves to unite these cells in forming, with connecting tissue, broad membranous plates, bundles,

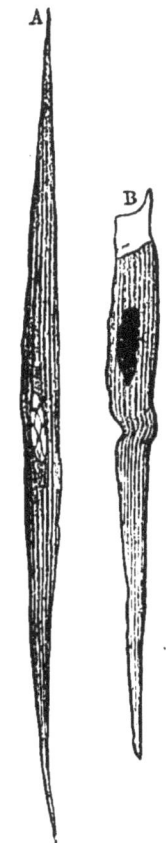

FIG. 52. NON-STRIATED MUSCLE. (SCHÄFER.)
A. Complete cell.
B. Broken cell.

plexuses, etc. It serves to afford contractility, especially to the organs of vegetative life.

Kill a good-sized frog by decapitation, and open the abdomen on the median line. Fill the bladder with air, after the introduction of a blow-pipe into the vent. Remove the inflated bladder with a single cut with the curved scissors, and place it in a saucer of water. Proceed to brush it, under the water, with two camel's-hair pencils, so as to remove all of the cells from the inner surface. It will bear vigorous rubbing with one of the brushes, holding it at the same time with the other. Transfer to alcohol for ten minutes, and afterward stain with hæmatoxylin and eosin. While in the oil, cut the tissue into small pieces, and mount flat in balsam. Examine with L and H.

Observe the bands of involuntary muscle crossing in various directions. You will distinguish between the muscle and the connective tissue cells by their nuclei.

STRIATED MUSCULAR TISSUE

A skeletal or striated muscle consists of cylindrical fibers, varying from 10μ to 100μ in diameter and 5 to 12 cm. long. These primitive fibers are supported by a delicate, transparent sheath— the *sarcolemma*. They are aggregated, forming *primitive fasciculi*, which are again united to form the larger bundles of a complete

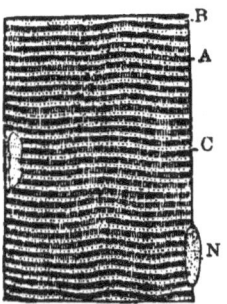

FIG. 53. PART OF A MUSCLE FIBER. (RANVIER.)
A. Dark disk.
B. Membrane of Krause.
C. Light disk.
N. Muscle nucleus.

muscle. The connective tissue uniting the primitive fibers is termed *endomysium;* while that uniting the primitive bundles is the *perimysium*.

The primitive muscular fibers exhibit marked cross striations with faint longitudinal markings, the former being produced by alternate dark and light spaces.

Under very high magnification each light band appears to be crossed by a dark line, called the *intermediate disk* or membrane of Krause. The dark band consists of rows of spindle-shaped bodies.

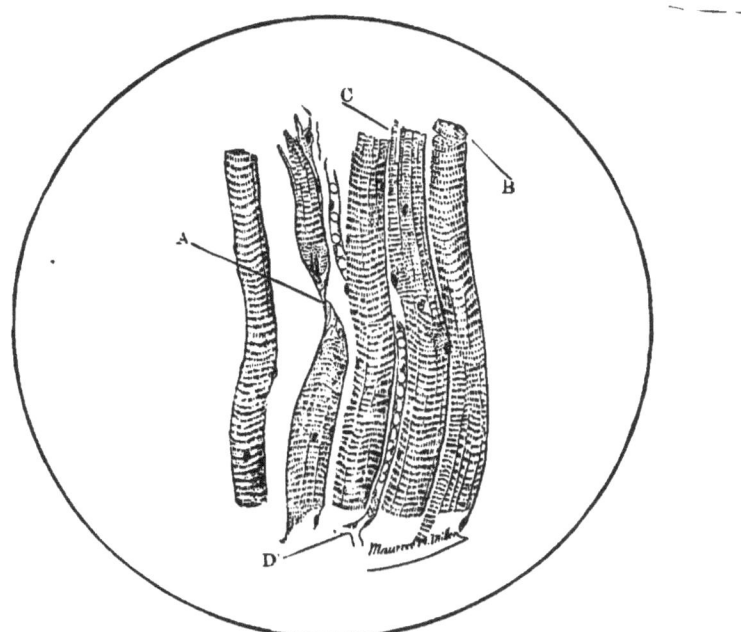

Fig. 54. Striated Muscular Fibers from the Tongue, Teased and Stained with Hæmatoxylin ($\times$ 400).

A. A fiber, with the muscle substance wanting, from stretching during the teasing, the sarcolemma alone remaining.
B. Partly separated disk of Bowman.
C. Ultimate fibrillæ.
D. A blood-capillary.

Those of the different dark bands are placed end to end, forming continuous elements. They constitute the *contractile fibrillæ*. The light bands correspond to the attenuated ends of the spindle-shaped bodies. Little knobs on the ends of the spindles make the dark line called the intermediate disk. The fibrillæ are held in bundles by a pale *sarcoplasm*. In transverse sections of muscle these bundles show as polygonal areas—*Cohnheim's fields*.

Macerate human muscle, preferably that from the tongue, in dilute chromic acid for twenty-four hours, wash, tease in water, cover, and focus with H. Fig. 54 was drawn from such a preparation.

The sarcolemma is best seen where the contractile substance has been broken. The muscle nuclei are seen at various points beneath the sarcolemma. Portions of a fiber have been split off transversely in places, indicating the *disk of Bowman*. The fibrillæ are indicated where the fiber has been split longitudinally during the teasing. The capillaries are arranged in a direction parallel to the fibers, with frequent transverse connections.

CARDIAC MUSCULAR FIBER

It presents the following characteristics:
1. The fibers are smaller than those of ordinary skeletal muscle.
2. They are striated both transversely and longitudinally.
3. They branch, forming frequent inosculations.

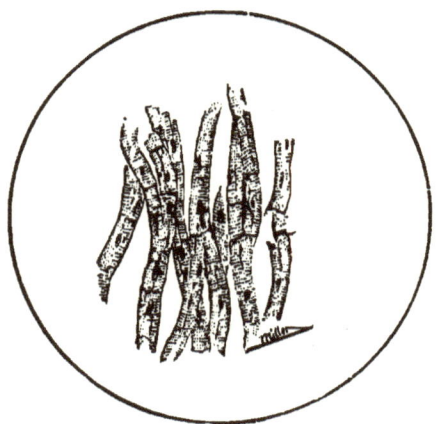

FIG. 55. TEASED CARDIAC MUSCULAR FIBERS.
Stained with hæmatoxylin. × 400 and reduced.

4. They are divided by distinct transverse lines into short cells.
5. Their nuclei are situated within the fiber.
6. They present no distinct sarcolemma.

Notice the groups of yellowish brown pigment-granules within the muscle-cells close to the nuclei. Fig. 55 has been drawn from fresh cardiac muscle, teased in normal salt solution and tinted with eosin.

NERVOUS TISSUES

Following the order given in the classification of tissues, the nervous tissues should be studied at this point. But in laboratory work it will be found more satisfactory to consider them in connection with the histology of the central nervous system (see page 216).

BLOOD

The human red blood-corpuscle is a flattened, bi-concave disk, circular in outline, and from 7 μ to 8 μ ($\frac{1}{3200}$ inch) in diameter. It

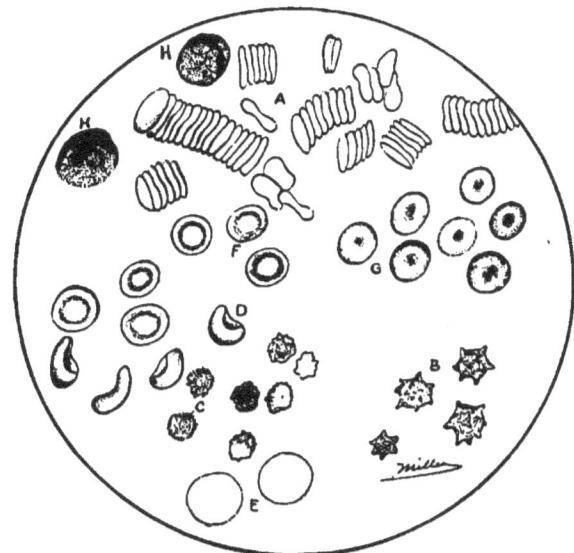

FIG. 56. CORPUSCULAR ELEMENTS OF HUMAN BLOOD ($\times$ 400).

 A. Colored corpuscles adhering by their sides—*rouleaux*.
 B. The same crenated.
 C. The same shrunken.
 D. The same having absorbed water.
 E. The same still more swollen.
 F. The same with the plane C D, Fig. 57, in focus.
 G. The same with the plane A B, Fig. 57 in focus.
 H. Colorless corpuscles.

presents a mass of protoplasm destitute, as far as the microscope shows, of nuclei, cell-wall, or any structure whatsoever.

Certain changes in form result, after removal from the circulation, viz.: 1. They may adhere by their broad surfaces forming columns. 2. From shrinkage they may become crenated. 3. Still further shrinkage produces the chestnut-burr appearance. 4. From absorption of water they may swell irregularly, obliterating the concavity of one side. 5. From continuous absorption they swell, forming spheres which are finally dissolved.

Wind a twisted handkerchief tightly around the left ring-finger,

FIG. 57. DIAGRAM OF A COLORED BLOOD-CORPUSCLE, SIDE VIEW, SHOWING THE BI-CONCAVITY. (The thickness is exaggerated.)

A, B. Upper plane, which, in focus, gives the appearance shown at G, Fig. 56.
C, D. Plane giving the appearance shown at F, Fig. 56.

prick the end with a clean needle, and squeeze a minute drop of blood on a slide, add a drop of salt solution, cover, and focus with H.

Observe: 1. That considerable variation in size of the red blood-corpuscles exists. 2. The color—a delicate straw tint. 3. That the concave centers of the corpuscles which lie flat can be made to appear alternately dark and light according to the focal adjustment. 4. That the concavity is also demonstrated as the corpuscles are turned over by the thermal currents.*

BLOOD-PLATES

Minute corpuscular elements in the blood, about one-fourth the size of the red disks, exist in the proportion of about one of the former to twenty of the latter. They are colorless ovoid disks, and are regarded by Osler as an essential factor in the coagulation of the blood.

Prick the thoroughly clean finger with a needle. Over the puncture place a drop of solution of osmic acid (one per cent.), and squeeze out a minute drop of blood, so that, as it flows, it is covered by the acid solution. This fixes the anatomical elements, providing against further change. The blood, as soon as drawn, must, with the acid, be immediately transferred to a slide and covered.

*The student is at this time advised to study the corpuscular elements of the blood of such animals as he may be able to command. The red corpuscles of mammals (excepting the camelidæ) do not vary in appearance from those of man, excepting in size. Those of birds, fishes, and reptiles are elliptical, with oval nuclei. Corpuscles of the blood of invertebrates are not colored.

To provide against evaporation, run a drop of sweet oil around the edge of the cover.

The blood-plates may be found, after careful search, bearing the relation to the red corpuscles seen in Fig. 58.

WHITE OR COLORLESS BLOOD-CORPUSCLES

The white blood-corpuscles are also called *leucocytes*. In fresh preparations they are seen to be perfectly colorless, nearly spherical cells, often slightly granular. The nucleus is distinguished with difficulty unless reagents are used. The leucocytes are not all of the same size. The larger ones are the more numerous. If they are watched carefully, the larger ones may be observed to change

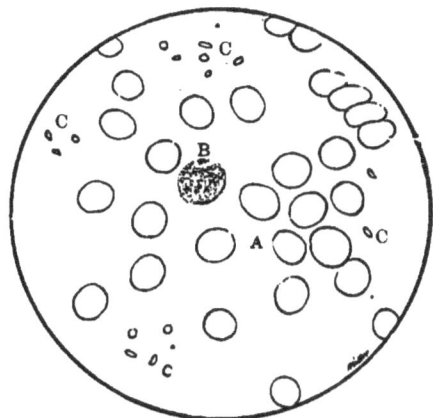

Fig. 58. Human Blood preserved with Osmic Acid.
A. Colored corpuscles.
B. Colorless corpuscle.
C, C, C. Groups of plaques. (× 400 and reduced.)

their shapes slowly. This movement is a property belonging to their protoplasm called amœboid movement.* Portions of the protoplasm are slowly extended outwards, making projections called pseudopodia, which may be drawn in again. By means of this movement the leucocytes can travel slowly from one part of the field to another. They are more active when the slide is warmed slightly.

---
*The Amœba is an extremely simple unicellular animal (Protozoön), which is found in the water of ponds. It is usually much larger than the white blood-corpuscle, and its movement is ordinarily very active and easily seen. The student should examine specimens of water and watch the movement of the Amœba.

The leucocytes, especially the larger ones, are of great importance in pathology in connection with inflammation and the formation of pus. The large leucocytes furnish the great majority of the cells in ordinary pus. The smaller leucocytes are about the size of the red corpuscles; the larger ones are about 13 µ in diameter. The leucocytes are very much less numerous than the red corpuscles. The ratio to the red corpuscles varies from 1 to 500 to 1 to 1,000. The leucocytes become more numerous a few hours after eating.

In order to study the leucocytes more carefully, they should be examined in dried and stained preparations, with an oil-immersion lens if possible. Square cover-glasses are used, which need to be clean and perfectly free from dust. They should be handled with forceps. Having cleaned and dried the skin of the finger, puncture it quickly with a clean, sharp needle, using no pressure. A drop of blood should be allowed to issue. Wipe away the first drop, and use the next, which should be no larger than a pin's head. Apply the surface of one cover-glass to the summit of the drop. Let this cover-glass fall on the other at the angle shown in Fig. 59. The blood is spread between the cover-glasses in a thin film. Quickly draw them apart, without lifting. The film of blood dries immediately. The object is to spread the blood on the cover-glass in a thin film within a few seconds after it leaves the capillaries, before coagulation or changes in the shapes of the cells can occur.

To fix the preparations they should be placed in a mixture of alcohol and ether (equal parts) for half an hour; or they may be subjected to dry heat (110° C.) preferably for half an hour or even longer.

There are many methods of staining. Much can be done with the ordinary hæmatoxylin and eosin stain. Very beautiful results can be obtained

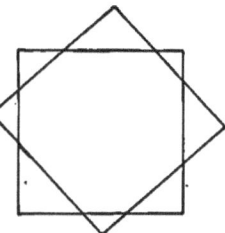

FIG. 59. MANNER OF PLACING COVER-GLASSES. (CABOT.)

with combinations of aniline dyes; for instance, eosin and methylene blue: one-half per cent. solution of eosin in sixty per cent. alcohol three minutes; wash; dry, by pressing between two pieces of filter paper; strong watery solution of methylene blue, one minute; wash; dry; balsam.

The very large nucleated red blood-corpuscles of the frog and newt should be stained in this manner.

To study the leucocytes of human blood after fixation, preferably with heat, use the Ehrlich tricolor stain (p. 32) for five minutes; wash, dry, mount in balsam. The red corpuscles are stained orange-yellow to brown. The nuclei of the leucocytes are stained green.

The Ehrlich method of staining shows us that the leucocytes are of several kinds. Some have large, round nuclei; others

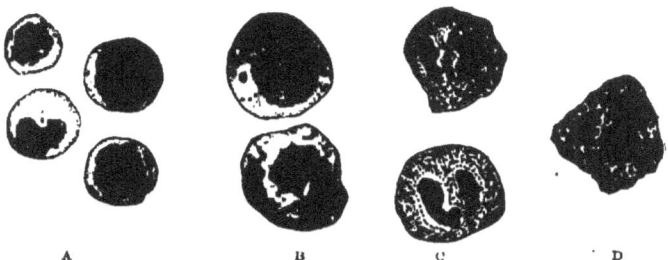

FIG. 60. VARIETIES OF LEUCOCYTES.
A. Small Lymphocytes.
B. Large Lymphocytes.
C. Polymorpho-nuclear Neutrophiles.
D. Eosinophile.

have nuclei that are distorted or that appear to be in several parts. Some have stained granules in their protoplasm; others have no granules. The nature of the granules and their affinities for the aniline dyes have been described in the chapter on staining. The granules in the leucocytes of man are of two sorts: (*a*) Neutrophile granules, very small and numerous dust-like granules of a reddish brown color; the leucocytes containing them look as though they had been sprinkled with red pepper. (*b*) Eosinophile granules—good sized, round, shining granules, not so numerous in the cells as the last.

These characters enable us to classify leucocytes as follows:

1. *Small lymphocytes*, which have a large, round nucleus and a thin band of protoplasm with no granules. They are the same as the lymphoid cells of the lymph-nodes and lymph from which they originate. They make twenty per cent. to thirty per cent. of all leucocytes.

2. *Large lymphocytes*, or large mononuclear leucocytes. They have a round or indented nucleus, and a considerable amount of protoplasm, without granules. Those with indented nuclei are often known as *transitional* forms. The large lymphocytes are not numerous—four to eight per cent.

3. *Polymorphonuclear neutrophiles* (often called polynuclear). The nuclei are much indented, or even seem to exist as several different parts. They contain immense numbers of fine neutrophile granules. They constitute the majority of leucocytes—sixty-two to seventy per cent.

4. *Eosinophiles*, which have polymorphous nuclei and eosino-

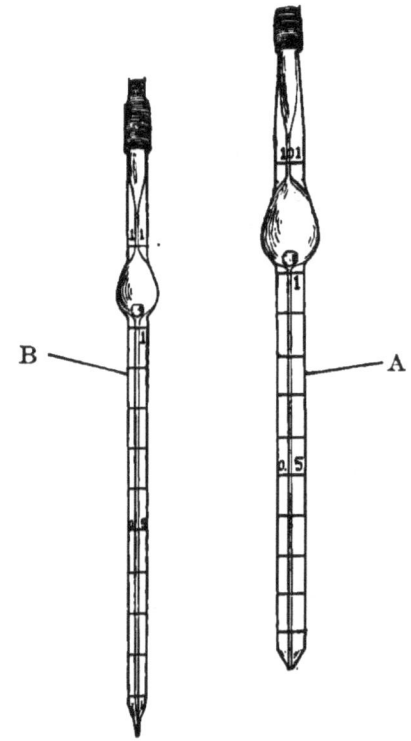

FIG. 61. MIXING PIPETTES. (CABOT.)
A. For red blood-corpuscles. It is the one referred to in the text.
B. For white blood-corpuscles, where the dilution is not so great. Weak acetic acid is used as a diluting fluid, which decolorizes the red corpuscles so that the white corpuscles alone are seen.

phile granules. They are of great importance in some of the diseases of the blood. The proportion of eosinophiles is from one-half to four per cent.*

---

*The percentages given are quoted from Cabot: Clinical Examination of the Blood, to which the student is referred for further information on this subject.

In disease the percentages above given are subject to much variation.

Cells that contain basophile granules are occasionally seen in normal human blood, as well as in disease. Their significance is not understood. They are not demonstrated by the Ehrlich tri-color stain. They are easily found in the blood of the Amphibia, by staining with basic dyes.

There is some reason for believing that the lymphocyte is the youngest form of leucocyte, and that the other varieties are developed from it in succession while circulating in the blood-channels.

ENUMERATION OF BLOOD-CORPUSCLES

The number of blood-corpuscles in a cubic millimeter of blood may be determined quite accurately by means of the hæmocytometer. To lessen the labor of counting, the blood is diluted with normal salt solution or Toison's fluid.*

The blood is drawn into the pipette, Fig. 61 A, to the point marked 0.5, or to 1.0, and then the fluid is drawn in till the bulb is filled to 101.0. With the finger on the end of the pipette it is shaken to mix the blood with the solution. Discarding the first drop, a small quantity is placed in the center of the small

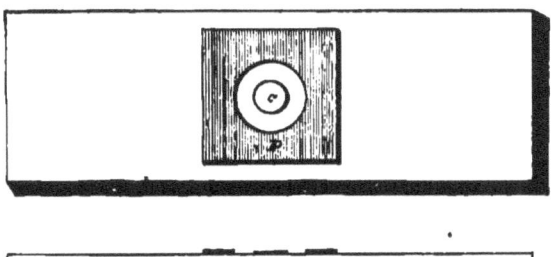

FIG. 62. PLATE AND RULED DISK OF THE HÆMOCYTOMETER.

disk in the middle of the slide, Fig. 62. The cover-glass, which goes with the instrument, is placed over the drop. Air bubbles are to be avoided, and no dust

---

*Toison's fluid—

Methyl violet, 5 B	.025 grams.
Sodium chloride	1. "
Sodium sulphate	8. "
Glycerin	30 c.c.
Water	160 c.c.

The leucocytes are stained faintly purple. The red blood-corpuscles retain their normal color and form.

particles or fluid should separate the cover-glass from the surface of the square of glass surrounding the disk.

The disk is depressed $\frac{1}{10}$ mm. below the upper surface of the square. It is also ruled into 400 squares $\frac{1}{20}$ mm. on each side. When the blood-corpuscles fall to the surface of the disk, as they do after a few minutes, the number counted in one square represents those present in $\frac{1}{4000}$ cu.mm. of the diluted blood. ($\frac{1}{20} \times \frac{1}{20} \times \frac{1}{10}$.) Every fifth square is crossed by a second ruled line, which encloses the small squares into groups of sixteens and assists in counting. Using the high-power, the corpuscles in a given number of squares in various parts of the plate are counted. In counting, discard the corpuscles that touch the lines on two sides (above and at the right), and

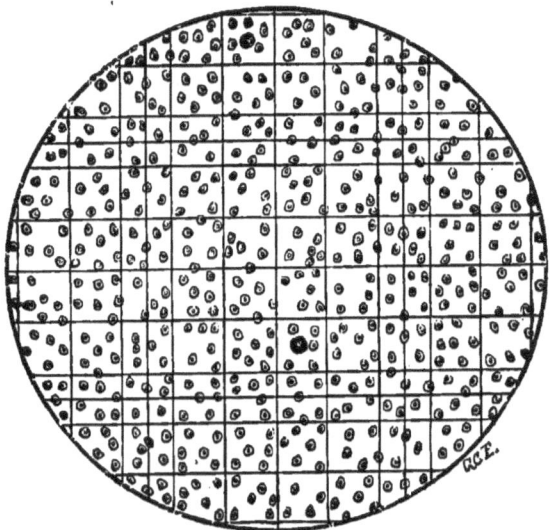

FIG. 63. APPEARANCE OF FIELD OF HÆMOCYTOMETER UNDER HIGH-POWER. (FREEBORN.)

count those that touch the other two (below and at the left); the average for one small square is thus taken. It is best to wipe away the first drop after a number of squares have been counted, replacing it with another, after thoroughly shaking the pipette. In all, count the number in about four hundred small squares. Multiply the average number for one square by 4,000 and by the dilution (100 or 200), and the number of red (or white) corpuscles in a cubic millimeter of blood is obtained.

The number of red blood-corpuscles in a cu.mm. of human blood is 5,000,000, or somewhat more, for men, and about half a million less for women. The normal number of white corpuscles is from 5,000 to 10,000 in a cu.mm.

HÆMOGLOBIN

The substance that gives to the red blood-corpuscles their characteristic color is hæmoglobin, which has the important function of being the oxygen-carrier of the corpuscle. The hæmoglobin of most mammals crystallizes in the form of rhombic prisms of a red color. That of the rat crystallizes quite readily.

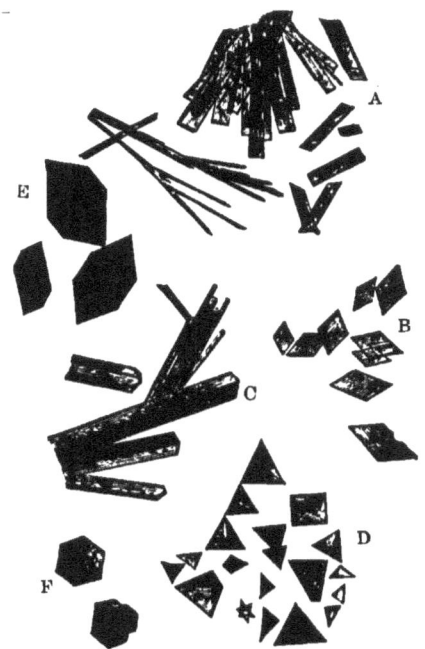

FIG. 64. CRYSTALS OF HÆMOGLOBIN. (RANVIER.)

A, B. Of man.
C. Of cat.
D. Of guinea pig.
E. Of hamster.
F. Of squirrel.

Hæmatoidin and hæmosiderin are substances derived from hæmoglobin often found in pathological tissues, as after hæmorrhages. Hæmosiderin contains iron, and occurs as yellow or brown granules. Hæmatoidin, which is the same as bilirubin, contains no iron, and occurs as granules or rhombic plates of a yellow to brown color.

HÆMIN CRYSTALS

Let a drop of blood dry on a slide. Add a few drops of glacial acetic acid, and heat over a flame until bubbles appear. Dry and mount in balsam. Dark brown rhombic prisms will be seen, which are crystals of hæmin, or the crystals of Teichmann. They are proof of the presence of blood, but do not indicate its source. They may be of importance in medico-legal cases.

FIBRIN

The delicate network of straight fibrin filaments is easily demonstrated by the method recommended by Gage. A large drop of blood is placed on a slide, and is covered with a cover-glass. The slide is laid on a piece of wet blotting paper, and covered with a saucer to prevent evaporation. After half an hour coagulation will have occurred. Draw a drop of water around the edge of the cover-glass, and float it carefully from the slide, endeavoring to keep the coagulum of fibrin on the cover-glass. Wash carefully in water, stain in hæmatoxylin and eosin; dry; mount in balsam. (Gage recommends mounting without balsam over a hard-rubber cell.)

EFFECT OF REAGENTS

Reagents produce characteristic changes in the blood-corpuscles. A strong saline solution leads to the formation of projections on

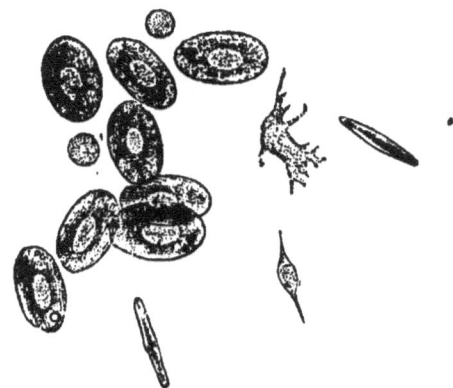

FIG. 65. BLOOD-CORPUSCLES OF FROG. (RANVIER.)

the red corpuscles, known as *crenation;* if sufficiently concentrated, the corpuscle becomes a shrunken, shapeless mass. Water causes the red corpuscles to swell; the hæmoglobin is finally dissolved out, leaving the colorless, barely visible outline of the stroma called the "ghost." After the addition of water the white corpuscles become spherical; their granules display the dancing "Brownian motion;" they swell, and finally burst. Weak acetic acid decolorizes the red corpuscles, and clears the granules of the white corpuscles so that their nuclei become visible. Weak solutions of tannic acid coagulate the coloring matter of the red corpuscles, which escapes from the cell, clinging as a minute particle to one edge.

DEVELOPMENT OF THE RED BLOOD-CORPUSCLES

The red blood-corpuscles of the mammalian embryo possess nuclei, and in this respect resemble those of birds, reptiles, amphibians, and fishes.*

The nucleated red corpuscles of the mammalian embryo and of the young forms of the lower vertebrates multiply by karyokinesis. At birth, however, in mammals the nucleated corpuscles are found to have been replaced by the ordinary, non-nucleated, discoidal forms. The origin of the non-nucleated corpuscles and the manner of their renewal throughout life are uncertain. It has been suggested that they are developed from leucocytes and also from the blood-plates, but both of these theories lack confirmation. It seems probable that the nucleated cells colored with hæmoglobin, found in the red marrow of the bones, are the most important source of the red corpuscles. According to Howell, the nucleated red corpuscles of the marrow lose their nuclei by extruding them.

---
*The red blood-corpuscles of the order of fishes known as Cyclostomi, of which the lamprey is a member, are circular, nucleated disks. Amphioxus has no red blood-corpuscles.

Part Third

ORGANS

THE SKIN

The skin consists of (1) the *epidermis* (or scarf skin), which everywhere covers and protects (2) the *derma* (corium or true skin).

The *epidermis* varies greatly in thickness in different locations; and in the thicker portions several layers may be differentiated. It is composed entirely of cells, while the derma is fibrous.

Epidermal Layers.
{
1. *Stratum Corneum*,
2. *Stratum Lucidum*,
} Horny Layer.
3. *Stratum Granulosum*,
4. *Stratum of Prickle Cells*,
5. *Stratum of Columnar Cells*.
} Malpighian Layer or Rete Mucosum.
} Epidermis.

The *stratum corneum* consists of old, exhausted, flattened, and desiccated cells, which are constantly falling from the entire surface of the body. Dandruff consists of impacted cells from this source. Those portions most frequently exposed to friction—*e. g.*, the palms of the hands and the soles of the feet—are protected by a corneous epidermal layer of great thickness.

The *stratum lucidum*, or clear layer, presents cells in form not unlike those in the preceding stratum; they are, however, translucent. This is properly a part of the previous stratum, is often absent, and frequently very difficult of demonstration. The stratum lucidum and stratum corneum owe their characteristic properties largely to the development in their cells of a substance called *keratin*.

The *stratum granulosum*, or granular layer, is composed of flattened cells containing opaque granules of *eleidin*, which is related to the keratin of the horny layers.

Immediately beneath the last-named layer, the cells become strikingly altered in form and appearance. The *prickle cells* are polygons or compressed spheroids, with large, oval nuclei, and minute, projecting spines. By means of these processes they are connected with one another.

The fifth and last (deepest) layer of the epidermis is composed of a single rank of elongated cells, placed with their long axes at a right angle to the surface of the skin. These cells contain the

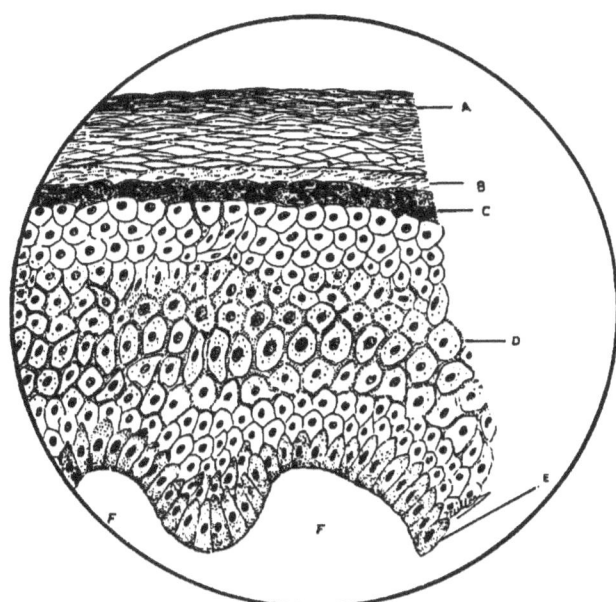

FIG. 66. VERTICAL SECTION OF THE EPIDERMIS FROM THE PALM OF THE HAND. STAINED WITH HÆMATOXYLIN AND EOSIN. (× 400.)

A. Stratum corneum.
B. Stratum lucidum.
C. Stratum granulosum.
D. Prickle cells of rete mucosum or rete Malpighii.
E. Stratum of elongated cells, the lower limit of the epidermis.
F, F. Indicate the position of two papillæ of the true skin or derma.

pigment which gives the hue peculiar to the skin of colored individuals.

The first two layers of the epidermis constitute, properly, the horny layer; while the remaining three strata compose the *rete mucosum* or *rete Malpighii*.

The *derma*, corium or true skin, is composed of dense, fibrillated connective tissue, so formed as to present minute elevations or papillæ over the entire surface of the body. These papillæ are covered with a basement membrane, and are protected from undue irritation by the epidermal layers.

The *subcutaneous cellular tissue* (upon which the true skin rests) consists of fibrillated connective tissue with elastic elements, from which strong interlacing bands are formed. These, in the deeper parts, form septa which support lobules of adipose tissue. These isolated collections of adipose tissue, when elongated and placed vertically to the surface, constitute the *fat-columns*.

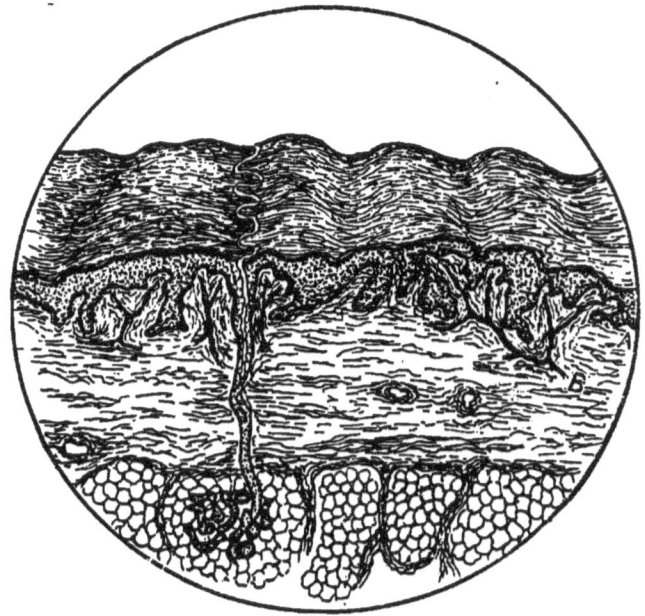

Fig. 67. Vertical Section Showing the Derma, or True Skin. Injected—Partly Diagrammatic.

A. Line of junction of derma with epidermis.
B. Capillaries distributed to papillæ.

The blood-vessels supplying the skin may be seen in vertical sections, in the subcutaneous tissue. Branches from these are sent to the papillæ, where they terminate in delicate, interlacing loops of capillaries.

Medullated nerves are also sent to the papillæ; and in certain locations they may be seen to terminate in tortuous structures— the *tactile corpuscles*. Varicose nerve-fibrils have been traced between the cells in the rete mucosum of the epidermis.

APPENDAGES OF THE SKIN

The appendages of the skin are the *hairs, sebaceous glands, sudoriferous glands,* and the *nails.*

THE HAIR

A hair, consisting of a root and shaft, is constructed from elongated, often pigmented cells, which are cemented together and overlapped with cell-plates, which form the *cuticle.* The central part of medullated hairs is composed of cubical cells and occasional minute air-bubbles.

The root penetrates the stratum corneum and (appearing to have pushed the rete mucosum before it) passes through the true skin and terminates in a bulb usually in the subcutaneous tissue, where it rests upon a papilla composed of an extremely delicate plexus of blood-capillaries.

The Hair-Follicle.—The root of the hair, in its passage to the papilla, is invested with sheaths derived from the skin. The hair,

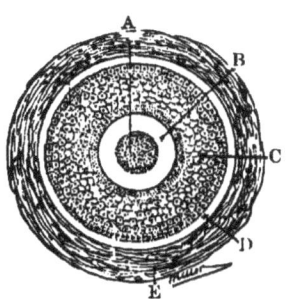

FIG. 68. TRANSVERSE SECTION OF HAIR AND HAIR-FOLLICLE. PARTLY DIAGRAMMATIC.

A. Medulla of hair.
B. Cortex of same.
C. Root-sheath.
D. Glassy membrane.
E. Fibrous wall of the follicle.

with its follicle, is indicated in transverse section in Fig. 68. A represents the *medulla,* and B the *cortex* of the hair. Outside the *root-sheath,* C, and derived from the rete mucosum of the epidermis, is a thin layer, the *glassy membrane,* D. This is projected from the basement membrane covering the surface of the corium, or true skin. The whole is surrounded by a *fibrous coat*, E, derived from the connective tissue of the derma.

A vertical section of the follicle is indicated in Fig. 69. A, B, and C represent the epidermal layers, which do not enter into its

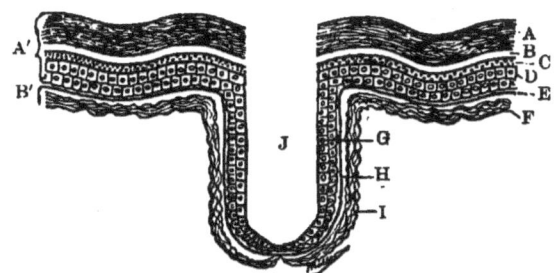

FIG. 69. DIAGRAM SHOWING MODE OF FORMATION OF HAIR-FOLLICLE

A'. Epidermal layers.
B'. Derma, or true skin.
A. Horny layer of epidermis.
B. Stratum lucidum.
C. Stratum granulosum.
  The three last mentioned form no part of the follicle.
D. Rete Malpighii. This will be seen projected into the depths of the true skin to form the root-sheath, G.
E. Hyaline membrane covering the derma. This is projected into the follicle, forming the glassy membrane, H.
F. Fibrous tissue of the derma, forming the fibrous sheath of the hair-follicle, I.
G. Root-sheath of the hair-follicle.
H. Glassy membrane of the follicle.
I. Fibrous sheath of the follicle.
J. The hair-follicle.

composition. The rete mucosum, D, forms the root-sheath at G. The basement membrane of the corium, E, forms the glassy membrane, H, while the connective tissue, F, constitutes the fibrous layer of the hair-follicle, J. The scales lining the hair-follicle are imbricated, and are directed downwards, fitting over the scales covering the surface of the hair, which are directed upwards, and also imbricated.

MUSCLES OF THE HAIR-FOLLICLES

Attached to the fibrous layer of each hair-follicle is a small band of involuntary or smooth muscular fiber—the *arrector pili*. This passes obliquely toward the surface of the skin; and when contraction takes place, the follicle and hair are elevated, producing the phenomenon known as goose-flesh.

SUDORIFEROUS OR SWEAT-GLANDS

A sweat-gland (Figs. 67 and 70) consists of a tube or duct which, from the opening upon the surface, passes in a spiral course through the several layers of the skin to the deeper part of the corium, where it becomes coiled in a bunch, as at D, Fig. 70. The coiled or glandular part of the tube is surrounded by a net-work of capillaries. At B the tube is seen in transverse section. The gland-tube, D, is provided with a wall of connective tissue and smooth or involuntary muscle, lined with conical

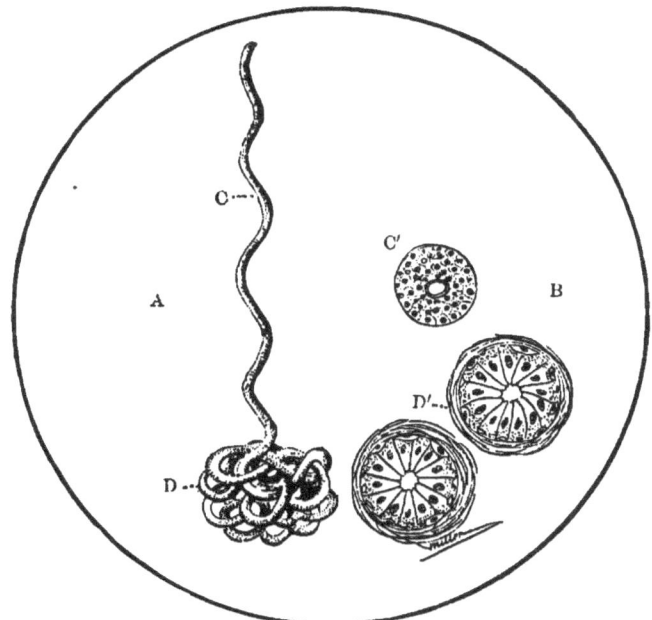

FIG. 70. SUDORIFEROUS TUBULAR GLAND.

A. Diagrammatic sweat-gland. C. Its duct. D. Coiled, glandular part.
B. The same, showing a transverse section of both parts (× 400). C'. The duct lined with several layers of cells. D'. The coiled glandular part lined with columnar cells in a single layer, resting on a basement membrane.

cells. The duct, C, is lined with granular epithelium covered with a thin cuticular membrane. Near the surface of the epidermis the lining cells disappear.

Krause estimated the number of sweat-glands at over two million.

SEBACEOUS GLANDS

These glands are little sacs or lobules, one or more of which open into each hair-follicle. These sacs are entirely filled with polyhedral epithelial cells (*vide* Fig. 71). At the neck of the gland

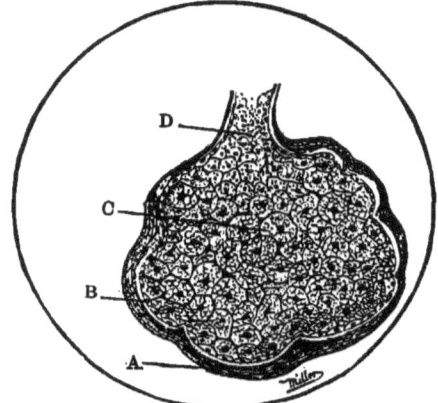

FIG. 71. SINGLE LOBULE OF A SEBACEOUS GLAND (× 400).
A. The fibrous wall of the sac.
B. Membrane propria.
C. Polyhedral cells filling the sac completely.
D. Fatty degeneration of the parenchyma at the neck of the gland, formation of *sebum*.

the cells become granular, fatty, and disintegrated, producing the *sebum*.

THE NAILS

The peculiar tissue of the nails corresponds to the stratum lucidum of the epidermis developed to an extreme degree. The nail rests upon a *nail-bed*, which represents the corium and the Malpighian layer of the epidermis. Minute longitudinal ridges take the place of papillæ. The root of the nails is imbedded in a part of the nail-bed called the *matrix*, from which its growth occurs.

PRACTICAL DEMONSTRATION

Remove the skin from the parts below as soon after death as practicable. Tissue may frequently be secured after surgical operations from stumps, etc. Dissect deeply, so as to preserve the subcutaneous tissue. Small cubes from the finger-tips, the palm of the hand, the scalp, and the groin may be hardened quickly in strong alcohol; and vertical sections should be made as soon as the tissue has become sufficiently firm. Stain with hæmatoxylin and eosin, and mount in balsam.

The structure of hairs may be best demonstrated by washing the soap from lather, after shaving, with several changes of water. When clean, decant the water and add alcohol. After twenty-four hours again decant and add oil of cloves. With a pipette carry a drop of the oil with the deposited hair-cuttings to a slide, remove as much of the oil as possible with slips of blotting-paper, and mount in balsam. Oblique, vertical, and transverse sections may be readily obtained by this method.

VERTICAL SECTION OF SKIN FROM THE GROIN
(*Vide* Fig. 72)

OBSERVE:

(L.)*

1. **The horny layer of the epidermis.** (The stratum lucidum will hardly be demonstrable on account of the thinness of the epidermis in this region.)

2. **The rete mucosum.** (The section from which the illustration has been drawn was taken from a negro, and the deep cells were pigmented.)

3. **The sharp line of demarcation between the epidermis and the true skin.**

4. **The papillae of the corium** or derma. (Note the absence of any sharp line dividing the corium and subcutaneous tissues.)

5. **The larger blood-vessels** of the subcutaneous region. (The *arteries* in transverse sections are plainly indicated by their prominent media, the appearance of the fenestrated membrane as a wavy yellowish line, and by the elliptical or circular outline. The **veins** are smaller, with thinner walls, and their outline is generally irregular. The smaller veins are commonly overlooked, on account of their lumen having become obliterated by contraction of the tissue in hardening.

6. **Coils and ducts of sweat-glands** in subcutaneous region. (The tubes are cut in various directions, and the whole is surrounded by dense fibrous tissue, forming a kind of capsule.)

7. The subcutaneous **collections of adipose tissue** beneath the last region. (The septa are dense and strong.)

8. (Having selected a vertical section of a hair-follicle:) (*a*) The **root of the contained hair.** (*b*) The **bulb and the hair-papilla.** (*c*) The **medulla of the hair.** (*d*) The **root-sheath** prolonged from the rete mucosum. (*e*) The **fibrous (outer) sheath.**

---

*Low-power—*i. e.*, from thirty to sixty diameters.

100   STUDENTS HISTOLOGY

9. The **sebaceous glands.** (The demonstration of the connection between the neck of the gland and the follicle will require a very favorable section.)

10. (Scattered through the corium and upper subcutaneous

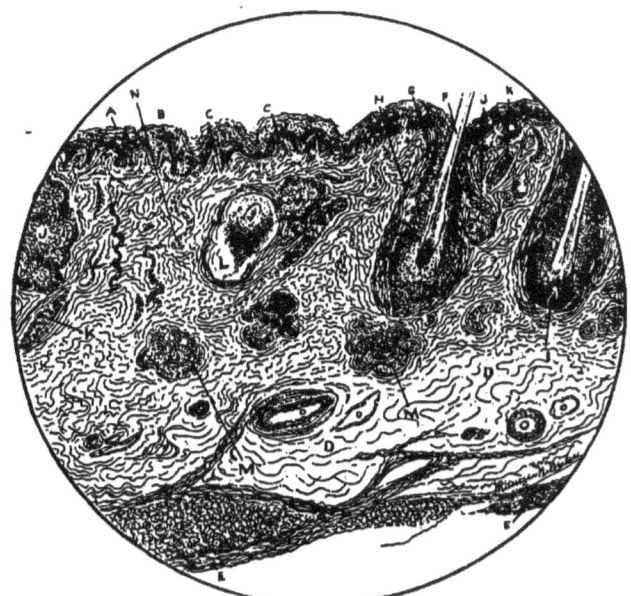

Fig. 72. VERTICAL SECTION OF SKIN FROM THE GROIN. STAINED WITH HÆMATOXYLIN AND EOSIN.

A. Epidermis.
B. Deep, elongated cells of the rete mucosum.
C. C. Papillæ of true skin.
D. D. Subcutaneous areolar tissue.
E. E. Collections of adipose tissue.
F. Shaft of hair (obliquely sectioned).
G. Root-sheath of hair.
H. Fibrous sheath of hair.
I. Hair-papilla (vertical section).
J. J. J. Portions of sebaceous glands (one on the extreme right of the cut is seen in connection with the hair-follicle.)
K. K. Arrectores pili.
L. Hair-follicle with contained shaft of hair in very oblique section.
M. M. Coils of sudoriferous glands.
N. Spiral duct of last.
O. O. Arteries of subcutaneous plane.

region:) (*a*) **Small portions of sebaceous glands.** (*b*) **Ducts of sudoriferous glands.** (*c*) **Oblique sections** at various angles of **hair-follicles.** (*d*) **Small vessels.**

11. **Arrector pili muscle.** (Nearly always to be found standing obliquely to the divided hair-follicle.

(**H.**)*

12. (If demonstrable:) (*a*) The **stratum lucidum.** (*b*) **Stratum granulosum.**

13. **The columnar cells of the rete,** next the corium.

14. (Where the tissue has been torn:) The **impacted cells of** the horny **epidermis.**

15. The **basement membrane** covering the corium.

16. **Capillaries of the papillae of the corium.** (These may be distinguished, when seen longitudinally, by tortuous lines of elongated and deeply stained nuclei belonging to their endothelium. Arterioles may be differentiated by their long muscle cells, the circular fibers lying transversely to the vessel.)

17. **The root-sheath of the hair-follicles.** (The cells composing the root-sheath vary in appearance, according to their position relatively to the hair; and this will enable you to demonstrate two layers, or an inner and an outer root-sheath.)

18. The **glassy membrane of the hair-follicle.** (Appearing simply as a clear space between the root-sheaths and the outer fibrous coat.)

19. The **intra-cellular network** in the large polyhedral epithelial cells of the sebaceous glands, and the minute **fat-globules** in the same.

20. The **nuclei of the fat-cells** in the adipose tissue. (They appear pressed to one side.)

21. **Medullated nerve-bundles** in transverse or oblique section.

*High-power—*i. e.*, from three hundred to four hundred diameters.

THE CIRCULATORY SYSTEM

THE HEART

The muscle of the heart has already been described. The muscle-cells are supported by a small amount of connective tissue, in which run the blood- and lymphatic vessels and nerve-fibers. Both medullated and non-medullated nerve-fibers are supplied to the heart, and minute ganglia also occur, especially in the auriculo-ventricular and the inter-ventricular furrows.

The PERICARDIUM is one of the great serous membranes. Its surface is covered by a single layer of flat endothelial cells, beneath which is a stratum of fibro-elastic connective tissue. This connective tissue is continuous with that running between the muscle-fibers. Underneath the pericardium there is usually more or less adipose tissue, especially along the course of the larger blood-vessels.

The ENDOCARDIUM is covered with a single layer of flat endothelial cells, which rest upon fibro-elastic connective tissue. The connective tissue is continuous with that supporting the muscle-fibers, and also joins with the lining of the blood-vessels that open into the heart.

The *valves* of the heart are duplications of the pericardium, containing abundant connective tissue. The muscle-fibers of the auricle extend a short distance into the auriculo-ventricular valves. A few blood-vessels may be present at the attached borders of the valves.

BLOOD-VESSELS

Blood-vessels include *arteries*, *arterioles*, *capillaries*, *venules*, and *veins*. They are all lined with flattened endothelial cells cemented by their edges; and their walls are constructed from non-striated muscular, yellow elastic, and fibrous connective tissues, in proportions varying according to the size and function of the vessel. Arteries are active, while the veins are comparatively passive agents in the circulation of the blood.

The large arteries are eminently elastic, from preponderance of yellow elastic tissue; while the arterioles are eminently contractile, from excess of muscular fiber.

Arteries possess three coats: the *intima* (internal), *media* (middle), and *adventitia* (external).

Fig. 73 represents a medium-sized typical artery. The intima, or internal coat, A, B, C, consists of a layer of flattened endothelial cells, which rest upon fibrous connective tissue, with a few elastic fibers. These structures are surrounded by a layer of elastic tissue, the elastic lamina or *fenestrated membrane*, which is the external limit of the intima. It appears in a transverse section as a wavy (from contraction of the media) shining line, and is an important element, from its relation to certain abnormalities of the blood-vessels. The media, D, consists of alternate layers of

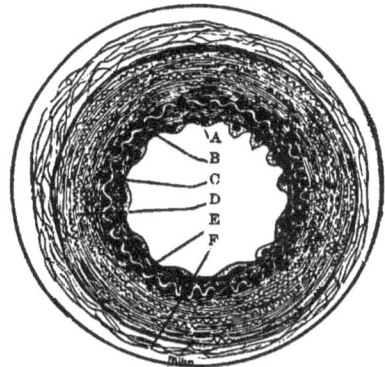

FIG. 73. TRANSVERSE SECTION OF A MEDIUM-SIZED ARTERY. PARTLY DIAGRAMMATIC.

A. The endothelial cells in profile.
B. Elastic and connective tissue supporting the endothelium.
C. The internal elastic lamina or fenestrated membrane. A, B, and C constitute the INTIMA of the artery.
D. The MEDIA. It consists of muscular and elastic tissues in alternating layers.
E. Points to one of the elastic layers.
F. The ADVENTITIA. Loose connective tissue, with few elastic fibers.

elastic and muscular tissue. The adventitia, F, is composed of fibrous connective tissue, containing some elastic elements.

As we approach the larger arteries, the muscular tissue diminishes in quantity and the elastic tissue is increased. On the other hand, the elastic element diminishes with preponderance of muscle as we approach the smaller arteries, until we meet the arterioles, the walls of which are made almost exclusively of involuntary muscular fibers, surrounding a layer of endothelial cells.

The walls of the capillaries consist of a single layer of flattened endothelial cells cemented by their edges. The union is not quite continuous, as minute openings are to be seen at irregular intervals.*

The AORTA has intima, media, and adventitia like the other arteries. The preponderance of the elastic tissue over the muscular, which is characteristic of large arteries, reaches its fullest development in the aorta. The intima is thick, and is not well marked off from the media. The great amount of elastic tissue in large arteries is connected with their function of converting the pulsating blood-current into a steady stream. The muscular fibers serve to control the calibers of arteries and the amount of blood flowing to any part.

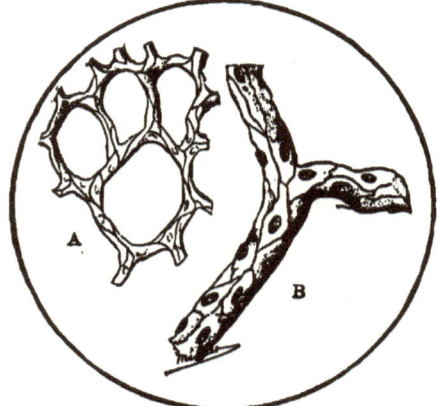

FIG. 74. ISOLATED BLOOD-CAPILLARIES.

A. Plexus from a pulmonary alveolus, stained with silver (× 350).
B. Capillary from omentum, stained with silver and hæmatoxylin (× 700).

In A the cells are outlined by the silver; while in B the nuclei in addition are brought out by the hæmatoxylin.

The walls of veins are much thinner than those of arteries. The intima presents an endothelial lining, but the line of demarcation between this coat and the media is often indistinct. The media contains muscular tissue but not much elastic tissue; and the adventitia, usually the most prominent of the three coats, is composed largely of fibrous connective tissue.

The valves of the veins are reduplications of the intima, having a semilunar form, and with the fibrous tissue well developed.

---

*It is probable that what appear to be openings between endothelial cells are, in fact, occupied by cement substance. In conditions of congestion and inflammation they become actual holes, and facilitate the migration of the leucocytes from the vessels by their amœboid movement, and permit, also, the diapedesis or escape of red corpuscles without rupture of blood-vessels.—(Klein.)

Vasa vasorum are small blood-vessels which serve to nourish the outer layers of the large arteries and veins.

PRACTICAL DEMONSTRATION

Sections of heart showing pericardium and endocardium, of aorta, of another large artery, and a large vein, should be studied and drawn. The small blood-vessels will be encountered in the various organs.

The DEVELOPMENT of capillaries is important because the larger vessels first appear in the embryo as capillaries, around whose endothelium the other coats later become differentiated from the neighboring mesoderm. Areas of mesodermic cells branch, unite with one another, and become hollowed out to form a system of channels. Part of the protoplasm and nuclei form nucleated red blood-corpuscles; part of them lie outside and constitute the wall of endothelial cells and their nuclei. Later in embryonic life and after birth the formation of capillaries is carried on in much the same way, and becomes of great importance in many pathological processes where new capillaries are required; for instance, in the healing of a wound. In these cases solid protoplasmic outgrowths are protruded from the endothelial cells of existing capillaries. These outgrowths lengthen by multiplication of the endothelial cells or by fusion with connective tissue cells. They branch and also unite with other similar outgrowths to form a network. Vacuoles appear in the middle of the processes, which enlarge, become confluent, and make channels through them, which open into the original capillaries. The protoplasm and nuclei of the solid sprouts form the endothelium of the new capillaries.

The newly forming capillaries may be studied in the thin tail of the young frog-tadpole. Select a tail with as little pigment as possible; harden in Flemming's chromic-acetic solution; wash thoroughly; stain with hæmatoxylin and eosin; alcohol; oil of cloves; balsam. Focus on a plane below the epithelium of the skin. The capillaries are narrow, dark bands, with nuclei; the large ones containing blood-corpuscles, the small ones solid and showing branches.

THE LYMPHATIC SYSTEM

The Lymphatic System is a circulatory apparatus of exceedingly complicated arrangement. It comprises:

1. A system of *irregular clefts and cavities which are of almost universal distribution in the more solid tissues, in the framework and parenchyma of organs, and around blood-vessels and viscera.* They are called *lymph-spaces* or *juice-canals*.

2. *Nódules of sponge-like tissue*, improperly called lymphatic *glands*.

3. *Channels of communication*, consisting of *capillaries and larger vessels called lymphatics*.

4. A *central reservoir*—the *receptaculum chyli*.

5. *Large efferent lymphatics*, by means of which the contents of the system are, eventually, poured into the blood, in both sides of the neck at the junction of the internal jugular and subclavian veins.

6. A fluid, *lymph*, containing numerous *lymphoid cells*, and various substances in solution.

The whole provides a channel for introducing formed and nutrient elements into the blood, and for conveying nutrition to the cells, as well as affording drainage for the tissues, the products of which are also emptied into the blood-vascular system, to be afterward eliminated by special organs.

The circulating lymph always passes in a direction toward the venous system. This current is established in some of the lower animals by means of distinct, pulsating, hollow organs, or lymph-hearts; but no corresponding structure exists in man, and the system becomes here subordinated to the blood-vascular apparatus.

In man, the maintenance of the lymph-flow is due largely to a negative pressure, consequent upon the connection between the termini of the lymph-vessels and the veins. Without doubt the pumping motion of the intestinal villi presents a factor in the establishment of a current in the lacteals toward the mesenteric vessels. The perivascular lymph receives an impetus with each cardiac systole. The muscular contractions of inspiration contribute motility to the contents of the diaphragmatic lymph-channels, in a direction against gravity. Indeed, the contractions of nearly every muscular fiber, whether skeletal or organic, lend their aid to lymph-propulsion.

The direction of the lymph-current is determined by valves which resemble somewhat those of the veins.

Cavities lined with so-called serous membranes may be considered as expanded lymph-channels.

LYMPH-CHANNELS

The larger and more regularly formed channels for lymph-circulation, such as the mesenteric and thoracic ducts, do not differ materially in structure from correspondingly sized veins. The irregular clefts in the interstices of fibrous tissues, serving as the primitive lymph-containing channels, will be repeatedly noticed in studying the various organs. Fig. 75, although purely dia-

FIG. 75. DIAGRAM. ARTERY IN TRANSVERSE SECTION, SHOWING THE PERIVASCULAR LYMPH-SPACE.

grammatic, will serve to show the relation of this system to the blood-vessels. A perivascular lymphatic channel is a sort of tubular investment of the blood-vessel, lined with flattened endothelium sending prolongations inward; these prolongations branch, and are finally in communication with a layer of cells covering the adventitia. In this manner, in close apposition to parts of the vascular system, a system of channels is provided, *within which the lymph may slowly percolate.* They are usually found about the arteries of the central nervous system.

The largest lymph-spaces in the human body are the cavities of the peritoneum and pleuræ. They are in connection one with the other, and with the lymphatic system generally; and the channels of communication between the great abdominal and thoracic lymphatic cavities are, perhaps, the most convenient and typical for demonstration.

LYMPHATIC VESSELS OF THE CENTRAL TENDON OF THE DIAPHRAGM (Figs. 76 and 77)

Practical Demonstration

This demonstration should be made with tissue from the rabbit, inasmuch as the slightest decomposition of the endothelium would be fatal to success.

A small (preferably white) rabbit should be quickly killed by decapitation, and immediately suspended by the hind legs, so as to thoroughly drain the body of blood. As soon as the blood has ceased dripping, open the thoracic cavity by slitting up the skin along the median line, pushing it to the sides and removing the sternum. In this operation work rapidly and avoid soiling the internal parts. Then with the fingers of one hand raise the lungs and heart from the diaphragm, and with a large camel's-hair brush proceed to quickly, and quite forcibly, pencil the white, glistening surface of the central diaphragmatic tendon, moistening the brush from time to time in the lymph of the pleural cavity. Should the quantity of fluid be small, add a little distilled or previously boiled and filtered water. The object of the brushing is to remove the endothelial cells which cover the surface, and which would otherwise hide the lymph-spaces. After the penciling, drain away the fluid and pour over the brushed surface a one-fifth per cent. solution of nitrate of silver.* Allow the silver solution to remain for twenty minutes in contact with the tissue, the body meanwhile being kept away from the bright sunlight; then pour off the solution, wash the surface twice with distilled water, and afterward allow water from the tap to flow over the parts for at least five minutes.

If you observe the directions carefully, the surface of the tendon will lose its original glistening appearance and become whitish and opaque.

The tendon, or such portion of it as you wish to preserve, may be cut out with the scissors after the washing, thrown into glycerin, and placed in the sunlight until the surface becomes brown. With the forceps tear off small pieces of the stained side, say one-half inch square, and examine in glycerin, or mount them permanently in the same medium.

The demonstration of the channels of the lymphatic system is based upon the following:

1. *Lymph-channels are always*, however small or irregular, *lined with flattened cells in a single layer—i. e.*, endothelium.

2. *The lining cells are cemented together with an albuminous substance.*

3. *Nitrate of silver combines with the cement, forming albuminate of silver, which becomes dark brown when exposed to light.*

---

* Water which has been well boiled in a clean vessel, and afterward carefully filtered, may generally be employed in histological work when distilled water is not available.

If you have been successful, the silver will have penetrated the tendon and mapped out the lymph-channels, *indicating an outline of every lining cell by means of a dark border.* Failure will result only from non-attention to cleanliness in the handling of the tissue, in which case the silver becomes deposited generally over the surface. The margins or outlines of the cells, it must be remembered, are stained with the silver. The nuclei may be demonstrated by after-staining with borax-carmine. The mounting may be done in balsam, although the elastic fibers, of which the matrix of the tendon is composed, will become stiff during immersion, and show a tendency to curl and contract. If glycerin be used after carmine staining, tissues should be washed thoroughly in water, subsequently to the acid alcohol, transferred to equal parts of glycerin and water, and allowed to remain for an hour, at least, before mounting.

CENTRAL TENDON OF THE DIAPHRAGM. SILVER-STAINING
(*Vide* Figs. 76 and 77)

OBSERVE:

(L.)

1. The **division of the specimen into dark and light areas.** (The dark areas represent the more solid portions of the tissue or the partitions between the channels, and the light spaces are the lymph-paths.)

2. The **lymph-paths**—the light spaces. (These show, with this amplification, as irregular, winding, and anastomosing courses, marked with very delicate lace-like tracery—the silver lines.)

3. **Valves of the lymph-paths.** (At points, the paths will be crossed by dark curved lines. These are imperfect valves, not unlike a single cusp of an aortic valve.)

(H.)

4. **Outlines of the cells** lining the larger excavations (lymph-paths) in the tissue. (Note that the cells are generally elongated in the direction of the lymph-path. The edges are frequently serrated.)

5. **Stomata**, minute openings at the junction of several cells.

6. The **construction of the valves.** (These are curved against the lymph-flow, and covered with cells like other parts of the channel. Note the change in form of the cells approaching and covering the valves.)

7. **Elastic fibers of the more solid parts of the tendon.**

8. **Lymph-capillaries.** (These will be seen in the partitions between the larger paths. In places they may be observed emptying into the paths, and again will appear as simple cavities, according to the manner sectioned.)

9. The **deeper capillaries.** (Careful focusing upon the portions of the tendon which appear most solid will reveal minute cell-lined channels or capillaries. The student must remember

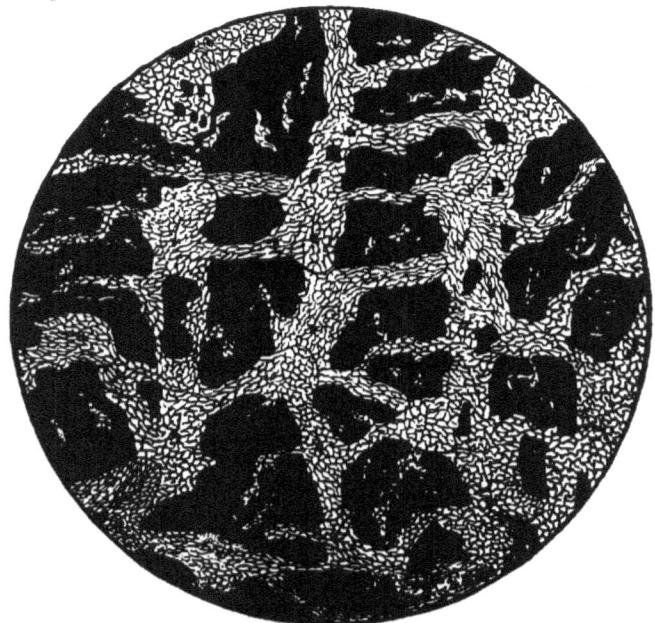

FIG 76. LYMPH-CHANNELS. CENTRAL TENDON OF DIAPHRAGM OF RABBIT. SILVER-STAINING (× 60).

The dark portions represent the more solid portions of the tissue.
The light areas are the lymph-channels; the direction of the flow is shown by the arrows.
The minute lines in the lymph-spaces are the silver-stained cement boundaries of the endothelial cells lining the channels.
The valves appear as curved lines in the lymph-spaces.

that we cannot penetrate tissues with the microscope to any considerable depth, but are restricted to nearly a single plane. If it were possible to penetrate with the eye the entire thickness of the tendon, we might trace the lymph-paths or channels from the abdominal to the thoracic surface.)

LYMPHATIC NODES

At numerous points along the course of lymphatic vessels they penetrate small nodules of so-called *lymphoid* or *adenoid tissue*, which have been termed lymphatic *glands*. They are frequently microscopic; others attain the size of a large pea. *They secrete nothing, hence are not glands.* They are somewhat sponge-like in structure, and the lymph filters slowly through them.

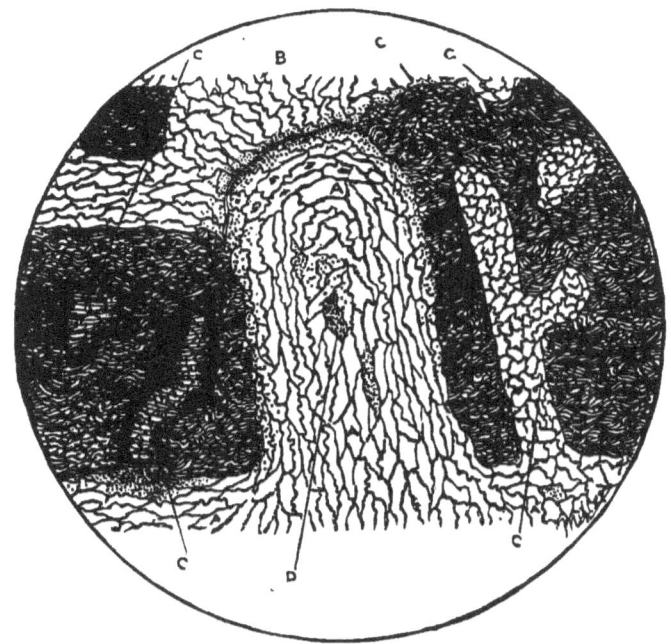

Fig. 77. A Small Portion of Specimen shown in Fig. 76, more Highly Magnified (× 350).

A, A, A. Large lymph-channel.
B. Valve in the course of last.
C, C, C. Lymph-capillaries in the more solid parts of the tendon.
D. Endothelial cells upon which a large amount of silver has deposited. Failure to follow the instructions for the staining frequently results in a like deposition of silver over the whole surface.

Most frequently several lymphatic channels enter one of these larger nodes, while perhaps only a single channel leaves it.

The histology of a lymph-node is not always easily comprehended by the student, and we have endeavored to make a diagram

(Fig. 78) which will simplify the matter somewhat. It is enveloped by a *capsule* of connective and involuntary muscular tissue, which sends *trabeculæ* into the body of the organ, and these branching posts support the structure as a framework. The interstices are quite small in the more central portion and larger toward the

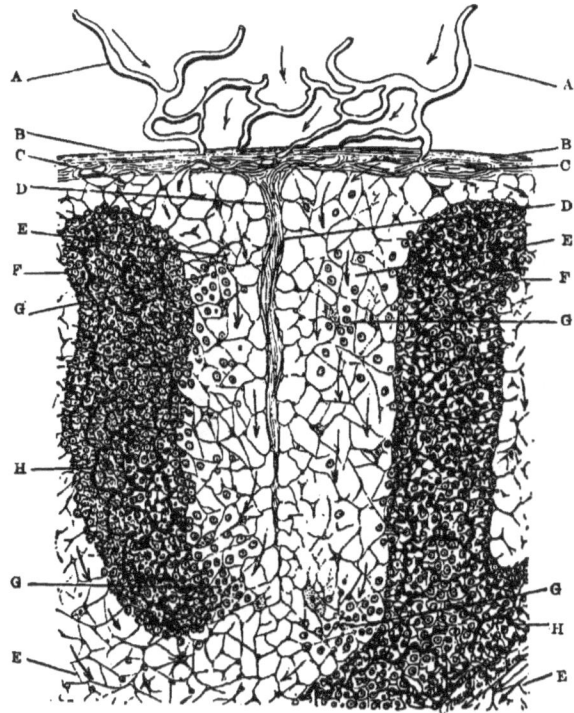

FIG. 78. DIAGRAM. PERIPHERAL PORTION OF A LYMPH-NODE.

A, A. Afferent lymph-vessels.
B. Capsule of the node, with lymph-spaces C, C.
D. Trabecula of connective tissue.
E, E, E. Lymph-path in the node.
F, F. Lymph-follicle of the cortex.
G, G, G. Lymphoid cells in the cell network of the paths.
H, H. Blood-capillaries of the follicles.
The arrows show course of lymph.

periphery; this has resulted in the application of the terms *medullary* and *cortical* to the respective parts. The nutrient blood-vessels are contained in the framework. *The compartments contain the structure peculiar to the lymphatic system*—viz., *lymphoid* or *adenoid tissue*.

Lymphoid or adenoid tissue consists of a mass of flattened cells, with numerous delicate fibrillar prolongations, which branch and anastomose so as to form an interwoven structure—*the adenoid reticulum*. Klein regards the cells as forming no essential part of the structure, but considers them as flattened plates attached to the fibrils. The meshes of the adenoid reticulum are in connection with the fibers of the trabeculæ, and, with the exception of the portion next the latter, are filled—crowded, in fact—with countless small, spherical lymphoid cells. Those portions of the tissue which contain the cells are termed the *follicles* of the cortex and *cords* of the medulla.

The *lymph-path* is the portion between the fibrous trabeculæ and the follicles and cords.

When we learn that the trabeculæ, follicles, cords, and lymph-paths pursue very tortuous and branching routes, we can appreciate the complexity of the organ as a whole.

The blood-vessel arrangement presents no anomalies. The small arterial trunks enter within the trabeculæ, finally break into capillaries which supply the follicles, cords, etc., and the blood is then collected by the venules for the efferent veins.

There is a depression at one side of the lymph-node called the *hilum*, where the arteries enter, and the veins and efferent lymphatic leave the lymph-node.

Small diffuse collections of adenoid tissue are to be seen in many organs. These do not differ essentially from the tissue just described, excepting that there is no definite arrangement of trabeculæ and lymph-paths, as in the compound lymph-node; the lymph simply *filters through the reticulum*, the same being a part of the lymph-channel system of the tissue in which the adenoid structure may occur.

PRACTICAL DEMONSTRATION

The *mesenteric lymphatic nodes* present the most typical structure, and may be obtained from the human subject, if fresh, although those from the dog are preferable, on account of the better condition of the tissue as usually secured.

The nodes should be sliced in half, placed in Müller for a week, and then hardened by two days immersion in strong alcohol.

Sections should be mounted, of two kinds, viz., those including the whole area of the node—which need not be very thin—for demonstration of the scheme or plan of structure, and exceedingly thin ones, even though they may include only a small part of the organ, for study of the details of the adenoid reticulum. The latter purpose will be subserved by shaking a number of thin

H

cuts in a test-tube with alcohol for a few minutes, and with considerable violence, even sacrificing most of the sections. The agitation will dislodge the lymph-cells, which otherwise would obscure the histology of the lymph-follicles and cords.

Stain deeply with hæmatoxylin and eosin, and mount the thicker sections in balsam, and those especially thin in glycerin.

SECTION OF MESENTERIC LYMPHATIC NODE
(Figs. 79 and 80)

OBSERVE:
(L.)
1. The **fibrous capsule.** (Note the elongated dots in the

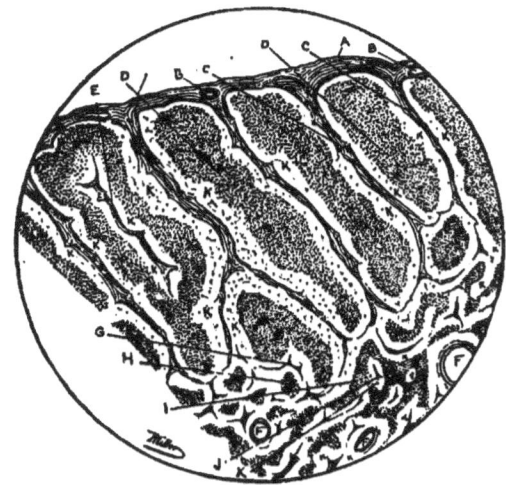

FIG. 79. VERTICAL SECTION OF A LYMPH-NODE FROM THE MESENTERY (× 60).

    A. Capsule of node.
    B. Lymph-spaces in the last.
    C, C. Trabeculæ, L. S. (Longitudinal section.)
    D, D. Follicle, L. S.
    E. Obliquely sectioned trabecula.
    F, F. Large blood-vessels of the central portion of the node.
    G. Trabecula in T. S. (Transverse section.)
    H. Medullary cord in T. S.
    I. Small and irregular cords of the center of the node.
    J. Obliquely sectioned trabecula of the center of the node.
    K, K, K. Lymph-paths.

deeper parts of the capsule—the **nuclei of the smooth muscular tissue, the thick-walled arteries, the lymph-spaces.**)

2. The **trabeculae.** (Trace these as they penetrate the organ, and observe that they frequently end abruptly, on account of hav-

ing curved, so as to leave the plane occupied by the section. The **trabeculae are not partitions,** like the interlobular pulmonary septa or the prolongations from the capsule of Glisson in the liver; they are not unlike rods or posts, making a framework and not producing alveoli. Find one divided transversely.)

3. The rounded **follicular** masses of lymphoid or adenoid tissue

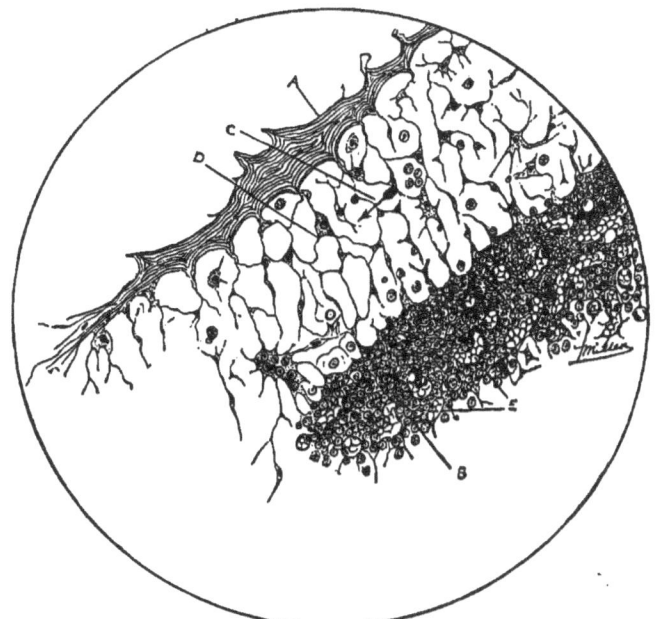

Fig. 80. Fragment of Section shown in Fig. 79. More Highly Magnified ($\times$ 350).

A. Trabecula.
B. Lymphoid cord.
C. Lymph-path.
D. Large branching cells of the lymph-path.
E. Capillaries of the cord.

in the cortex, and the **cord-like** masses of it in the medulla. (They are recognized as granular areas between the trabeculæ.)

4. The **lymph-paths.** (These can be appreciated by remembering that the follicles and cords do not entirely fill the spaces between the trabeculæ, and that the area between the two—*i.e.*, outside the cords—is the more open in texture, and contains the filtering lymph. They are more distinct in the cortex.)

(H.)

5. The **histology of the capsule.** (*a*) The closely united connective tissue with the scattering **elastic fibers** of the **external layer.** (*b*) The **smooth muscle of the deeper portions.** (*c*) Sections of **arteries.** (These may be of considerable size.) (*d*) The **lymph-spaces.** (The differentiation is by the flattened **endothelial** cells of spaces which otherwise would be supposed mere rifts in the tissue, inasmuch as no definite or special wall can be detected.)

6. The **structural elements of the trabeculae.** (They are similar to those of the capsule, excepting the elastic element, which cannot here be demonstrated. Note the variously sectioned small arteries.)

7. The **follicles** of the cortex and **lymphoid cords** of the medulla. (In the thicker section, the field will be completely crowded with lymphoid cells. Select a thin field and observe: (*a*) The **lymphoid cells.** (These will be found varying in size from a very small red blood-disc to that of a large white corpuscle; the nucleus is usually large, single or otherwise, while the protoplasm is often scanty.) (*b*) The **branching endothelial cells.** (*c*) The delicate fibrillæ of the **adenoid reticulum.** (You may endeavor to determine whether this reticulum exists as an offshoot of the endothelial cells, or whether the latter are simply adherent to the broadened plates of the former.)

8. The **reticulum of the lymph-paths.** (Observe that this is precisely like the reticulum of the follicles, as demonstrable after shaking out most of the lymph-corpuscles of the last.) (*a*) The **connection between the fibrillae of the paths and those of the trabeculae.**

9. **Capillaries of the paths and cords.** (These will be recognizable only by the regular succession of the contained red blood-corpuscles.)

THE SPLEEN

The spleen presents no regular subdivision of parts which may be studied separately and combined afterward, as we are able to do with organs like the lung, liver, etc. The spleen is a *ductless* organ or so-called gland, and the plan or scheme may, perhaps, be best comprehended by following the blood distribution.

The splenic artery enters the organ at the hilum, supported by a considerable amount of connective tissue, and rapidly breaks into smaller branches, from which the arterioles leave at right angles. The arterioles quickly merge into capillaries, which form plexuses

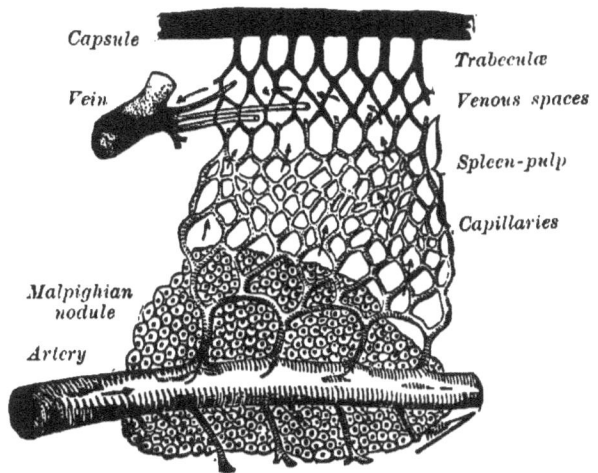

FIG. 81. DIAGRAM. SHOWING THE COURSE OF BLOOD IN THE SPLEEN.

throughout the different portions of the organ. Here we meet with an anomalous structure.

The capillaries, instead of uniting to form venules, as in the usual vascular plan, *empty their contents into small chambers or sponge-like cavities—the venous spaces.* The blood, after filtering through the venous interstices, is collected in larger, irregular, vein-like channels, which finally conduct the blood into the veins proper and out of the spleen. The tissue containing this vascular arrangement is called *splenic pulp*.

The fibrous capsule which envelops the spleen sends trabeculæ within, which form a framework; and from this fibrils are sent off, which branch, broaden, and inosculate to form the supporting framework of the pulp.

The arteries are frequently surrounded by nodules of lymphoid (or adenoid) tissue, sometimes globular, more frequently considerably elongated, and following the vessel for a considerable distance. These nodules are called *Malpighian bodies*. They bear no resemblance to the similarly named structures in the kidney, excepting, perhaps, when seen in transverse sections by the naked eye.

The spleen will thus be seen to consist of fibrous trabeculated *framework*, the *pulp*, *blood-vessels*, and more or less isolated nodules of *lymphoid tissue*.

PRACTICAL DEMONSTRATION

The organ must be perfectly fresh. If human tissue cannot be obtained in good condition, recourse may be had to the ox, which will provide an excellent substitute. The small supernumerary spleens, not infrequently found during *post-mortem* work, are most desirable, as sections can be easily made through the entire organ.

Pieces of tissue a centimeter thick, including a portion of the capsule, should be hardened as directed for lymph-nodes. Sections are easily made without the microtome, as the mass is very firm; they should be thin and stained with borax-carmine or hæmatoxylin, and mounted in balsam.

SECTION OF HUMAN SPLEEN, CUT AT A RIGHT ANGLE TO AND INCLUDING THE CAPSULE. (Fig. 82)

OBSERVE:

(L.)

1. The **fibrous capsule**. The clear, **translucent appearance** of its *elastic tissue* and the elongated nuclei of the smooth muscle-cells. (The capsule not infrequently becomes considerably thickened in the human subject, and the development occurs irregularly, sometimes in the form of minute nodules.)

2. The **trabeculae**. (The depth to which they may be traced will depend largely upon the direction of the section.) (*a*) That **these are not bands**, but bundles, more or less circular in transverse section. (*b*) Their **irregular course**, quickly after leaving **the surface**. (*c*) That occasionally a **small artery** may be found within them, though they are usually destitute of large vessels. (*d*) The elongated nuclei of the **muscular fibers** forming part of the trabeculæ.

3. **The large blood-vessels.** (*a*) The **arteries** more frequent than **veins.** (*b*) Their very prominent **adventitia.** (*c*) Their **tortuous** course.

4. The **lymphoid (or adenoid tissue).** (This you will be enabled to recognize by the great number of lymphoid cells of the

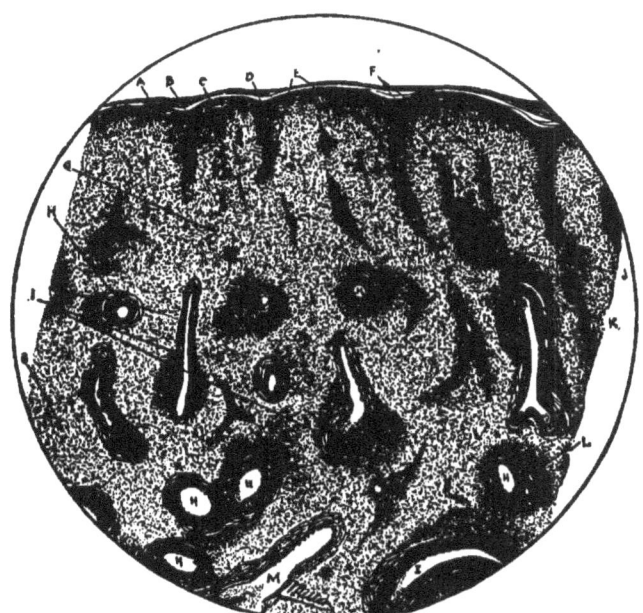

FIG. 82. SECTION OF THE SPLEEN ($\times$ 60).

A. Elastic portion of the capsule.
B. Lymph-spaces of last.
C. Involuntary muscular portion of capsule.
D. Deeply pigmented portions of capsule.
E, E. Trabeculæ from C.
F. Trabeculæ in oblique section.
G, G. The splenic pulp.
H, H. Large arteries in T. S.
I. Arteries in L. S.
J. Adenoid nodule, not connected with an artery.
K. Adenoid nodule—Malpighian body—along course of artery.
L. Adenoid nodule in T. S.
M. Vein.

adenoid structure, the nuclei of which become stained very deeply blue with hæmatoxylin, giving a very distinct differentiation. At this point examine every part of the specimen closely, and endeavor to detect even the most minute collection of this tissue.) (*a*) Around arteries, constituting the so-called **Malpighian bodies.**

(*b*) **Transverse sections of Malpighian bodies,** noting that the vessel is seldom in the center of the nodule. (*c*) Nearly longitudinal section of **Malpighian nodules,** observing that the lymphoid tissue usually follows or surrounds the artery for a short distance only. In many of the lower animals the Malpighian bodies are more sharply marked off from the splenic pulp than in man.

5. The **Splenic pulp.** (This will be found in those portions of the section not occupied by structures previously demonstrated; and will be determined by its light color. Review the whole area, and endeavor to differentiate every portion of the lymphoid and pulp-tissue. The staining will have been your principal guide thus far, the pulp-elements appearing in strong contrast by their pink eosin color.)

(H.)

6. The **structural elements of the capsule.** (*a*) The **numerous minute lymph-spaces** and the **imperfect vascular supply.** (*b*) The nuclei of the **peritoneal cell covering.** (This presupposes that the section has been selected so as to include the peritoneal investment.) (*c*) The abundant and closely packed connective tissue. (*d*) The **muscle-nuclei.** (*e*) Cells containing granular yellow **pigment.** (The quantity varies largely with different specimens.)

7. The **Malpighian nodules.** (*a*) The **arterioles**—very small and apt to escape attention unless filled with blood-corpuscles. (*b*) Their **reticulum.** (This will be difficult of satisfactory demonstration, unless the section is thin.)

8. The **elements of the pulp.** (*a*) **Large flattened cells,** the branches forming the meshwork of venous channels. (*b*) **Red blood-corpuscles.** (Very numerous, and often broken and distorted.) (*c*) **Blood-pigment.** (*d*) **White blood-corpuscles.** (Some of them may contain granules of pigment, which are composed of the derivatives resulting from the disintegration of hæmoglobin. The destruction of worn-out red blood-corpuscles probably is one of the most important functions of the spleen.) (*e*) **Large multi-nucleated cells,** (which are commoner in the spleens of young animals).

THYMUS BODY

The thymus body (frequently and improperly called a gland) is an adjunct to the lymphatic system of—in man—fœtal and infantile life, disappearing by an atrophic process at or before the age of puberty.

It is enveloped by a fibrous capsule, partitions from which subdivide the organ into lobes and lobules. The lobules are generally subdivided into follicles, which are irregularly sized and shaped, while tending to an ovoid form.

It is in connection with the general lymphatic system by efferent vessels which emerge from the hili of the lobes—the lymph having meanwhile traversed the mesh-like structure of adenoid tissue composing the follicles.

The blood-vascular system is in the form of a nutritive supply; the larger vessels occupying the fibrous framework, and sending branches into the follicles. The capillary plexuses are more abundant in the peripheral portion of the follicles. The blood is collected in the venous channels of the central or medullary area, and emerges from the organ by the veins which accompany the efferent lymphatics.

PRACTICAL DEMONSTRATION

The organ should be obtained from a still-born infant, divided in small pieces, and hardened rapidly in strong alcohol. Sections may include an entire lobe, and be stained with hæmatoxylin and eosin.

SECTION OF THE THYMUS BODY FROM AN INFANT AFTER DEATH ON THE SIXTEENTH DAY

OBSERVE:

(L.)

1. The **fibrous capsule**.
2. **Division by prolongations** of 1 into somewhat spherical lobes.
3. **Subdivision** of 2 **into lobules**.
4. **Subdivision** of 3 **into follicles**. (Note that these are not uniformly outlined by the connective tissue.)
5. The **subdivision of the follicles** into an outer, deeply stained **cortex**, which completely surrounds a light center, the **medulla**.
6. The **larger lymph-spaces** and **arteries** of the capsular and trabecular tissue.

(H.)

7. The **cortex of the follicles**. (*a*) The numerous deeply-

stained **lymph-corpuscles**. (*b*) The network of the lymphoid (or **adenoid**) **tissue**. (This will be greatly obscured by the lymphoid cells.) (*c*) The **blood-capillaries**. Only recognized by the contained corpuscles. (*d*) Minute **trabeculae** of the connective tissue projected from the capsule.

8. The **medulla of the follicles**. (*a*) The **sparsity of lymph-corpuscles** as compared with the cortical portions. (*b*) **Large**

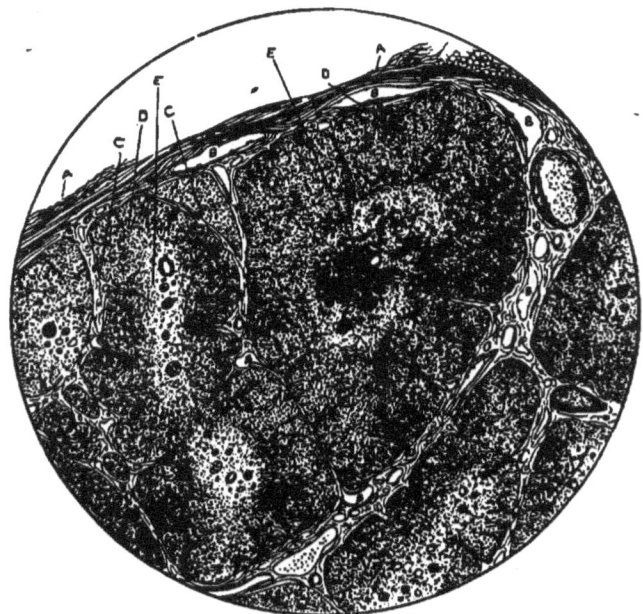

Fig. 83.—Section of a Portion of the Thymus Body from a Child Sixteen Days after Birth (× 60).

- A. A. Capsule which divides the organ into lobes. Portions of six lobes are visible in the section.
- B. B. Lymph-spaces.
- C. C. Trabeculæ dividing the lobes into imperfect lobules.
- D. D. Subdivisions of the last into follicles.
- E. E. Central light portion of the lobules.

mononucleated cells. (*c*) Still **larger multinucleated cells**. (*d*) Larger—though varying in size—spherical bodies, **Hassall's corpuscles**. (These are composed of epithelial cells, arranged concentrically, and are unlike any other structure found in the normal tissues of the body. They resemble the smaller "cell-nests" of epithelioma. The corpuscles of Hassall are the remains of the epithelial structure, which makes up the bulk of the thymus body in its early stages.) (*e*) Small thin-walled **venules**.

THE RESPIRATORY ORGANS

The *larynx* and *trachea* have the same general structure as the larger bronchial tubes. The epithelium is stratified columnar and ciliated, except that covering the surfaces of the epiglottis and that of the upper part of the larynx, which is stratified squamous. The cartilages are hyaline, except those of the epiglottis, part of the arytenoids, and the cartilages of Wrisberg and Santorini, which are yellow elastic cartilage.

THE LUNGS

At the root of each lung the large primary bronchus enters, and divides into two branches, which also divide and branch repeatedly until terminal or capillary bronchial tubes are formed, which are one-fourth to one-eighth of a millimeter in diameter.

A typical bronchial tube (Fig. 84) presents four coats, as follows:

1. *Epithelial.*
2. *Internal fibrous or mucosa.*
3. *Muscular or muscularis mucosæ.*
4. *External fibrous or submucosa.*

The *lining epithelium* is composed of cylindrical cells, provided on their free extremities with delicate hair-like appendages—the *cilia*. Between the pointed, attached ends of the ciliated cells, small ovoid cells are wedged, and the whole rests upon a layer of round cells. The epithelium pursues a wavy course, so that the lumen of a tube appears stellate rather than circular in transverse section. This greatly increases the extent of surface.

The *internal fibrous coat* or mucosa is composed of a small amount of connective tissue, which, just beneath or outside the epithelium, sustains collections of *lymphoid* (or *adenoid*) *tissue*. In the pig, a considerable quantity of *yellow elastic tissue* is found in the mucosa outside the lymphoid tissue, but the amount is smaller in man. The fibers are, for the most part, disposed longitudinally. Many *nutrient vessels* from the bronchial artery, capillaries, venules, and lymph-spaces are also found in this coat.

The *muscular coat*—muscularis mucosæ—does not differ from the same layer in other mucous membranes. Its thickness varies in proportion to the size of the bronchus, the smaller tubes possessing relatively the thicker walls. The fibers pass circularly, and are of the non-striated or involuntary variety.

The *external coat*, or submucosa, is largely composed of loose connective tissue, the fibers being mostly arranged circularly. A few delicate elastic fibers run longitudinally. The external fibers, like those of all tubes, ducts, and vessels, are for the purpose of establishing connection with the organ or part traversed; so that it is often difficult to demonstrate the exact external limit of a bronchus. This coat is liberally supplied with nutrient branches from the bronchial artery.

The elasticity and strength of the larger and medium-sized

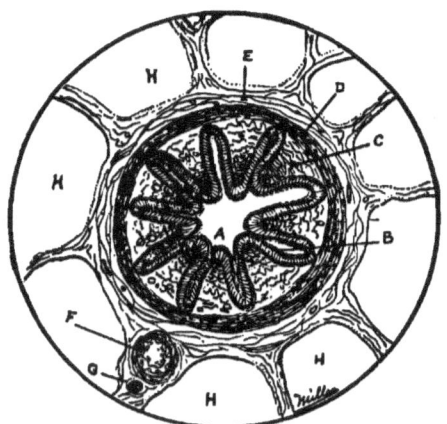

FIG. 84. TRANSVERSE SECTION OF A PORTION OF HUMAN LUNG, SHOWING A SMALL BRONCHIAL TUBE ($\times$ 60). STAINED WITH HÆMATOXYLIN.

    A. Lumen of bronchus.
    B. Ciliated columnar epithelium.
    C. Internal fibrous layer—*mucosa*.
    D. Muscular coat.
    E. External fibrous layer—*submucosa*.
    F. Pulmonary artery.
    G. Nerve.
    H, H, H. Pulmonary alveoli surrounding bronchus.

bronchial tubes are greatly increased by the presence of *cartilage* in the form of *plates*, which are imbedded in the external coat. They are not uniform in size, neither are they placed regularly. They frequently overlap one another, and two or three may be superposed. As the tubes become reduced in size the plates become diminished in frequency—disappearing altogether when a diameter of about one millimeter has been reached. The cartilage is of the hyaline variety; and each plate is covered with a dense fibrous coat, the *perichondrium*, which unites it with contiguous parts.

The principal bronchi are provided with a great number of *mucous glands*, which are located in the external coat or submucosa. They are simple, coiled tubular glands, commencing on the inner surface, penetrating the mucosa and muscularis mucosæ, and terminating in the submucosa, generally within the cartilage, where they are coiled in short, close turns in sections resembling somewhat the larger sweat-glands of the skin. The ciliated epithelium of the bronchial tube is continued down the beginning of the tube for a short distance, after which the cells are shortened, and lose their cilia. The coiled gland-part of the tube is lined with conical cells, which are so large as to leave the lumen very small. Sometimes, and especially in the aged, an ampulliform dilatation of the tube may be seen during its passage through the mucosa.

The description just given will apply to large and medium-sized bronchial tubes. Very important changes take place as we pass to the terminal tubes.

As the tubes decrease in size, the first coat to diminish in thickness is the outer, or submucosa. We have already alluded to the disappearance of the cartilage, and the mucous glands are lost at about the same time. The outer coat becomes, in the small bronchial tubes, so thin as to be no longer distinctly demonstrable. The muscular coat is the last to disappear. It remains a prominent feature of the tube as long as separate coats can be distinguished. The epithelial cells lining the tubes toward the termini become shortened, and, getting lower and lower, at last result in cuboidal cells, without cilia.

The walls of *terminal bronchial tubes* (diameter one-fourth to one-eighth of a millimeter) are composed of a slight amount of connective tissue in which an occasional non-striated muscle-cell and yellow elastic fiber can be distinguished. They are lined with cuboidal or a few flat cells. No definite layers are distinguishable in these bronchial tubes. In a transverse section the lumen would appear circular.

PRACTICAL DEMONSTRATION

The histology of the bronchi can be studied to best advantage by using tissue from a freshly killed pig, cat, or dog. Short pieces of tubes, about one centimeter in diameter, from which most of the lung-substance has been cut away, should be hardened quickly in strong alcohol. Transverse sections can be made free-hand, or the tissue may be infiltrated with paraffin or celloidin, and cut with the microtome. Stain with hæmatoxylin and eosin, and mount in balsam.

TRANSVERSE SECTION OF PORTION OF BRONCHUS OF PIG
(Fig. 85)

OBSERVE:

(L.)

1. The **epithelial lining**: (*a*) The **wavy course**. (*b*) Regions occupied by **beaker or goblet cells**. (The letter E in the drawing leads to such a group. (*c*) The number of **nuclei**, indicating the presence of more than a single layer of cells.

2. The **mucosa**. (*a*) Deeply stained blue nuclei of the **lymphoid (or adenoid) tissue** just beneath the epithelium. (*b*) Pink portion of the region below the lymphoid tissue. (The longitudinal **elastic fibers** cut transversely.) (*c*) **Blood-vessels**.

3. The **muscular coat**. (*a*) Apparent solution of continuity in places caused by **tubes of mucous glands**. (*b*) **The absence of large vessels in this coat**.

4. The **external layer**. (*a*) Its extent. (It includes the remainder of the section.) (*b*) **Large cartilage plates**, C, stained blue. (*c*) **Cartilage-cells**. (Note their differing forms and disposition in rows next the surfaces of the plates.) (*d*) **Perichondrium** stained pink. (*e*) **Mucous gland-coils**. (They are usually between the cartilage and the muscular coat.) (*f*) Section of **bronchial arteries and veins**. (*g*) Collections of **adipose tissue** on the outer surface. (*h*) Portion or whole of **pulmonary artery** and **medullated nerve-trunks** outside of and accompanying the bronchus.

(H.)

5. **Epithelial lining**. (*a*) **Cilia** of columnar cells. (*b*) The **ovoid cells** between the tapering columnar cells. (*c*) The "**basement membrane**," upon which the columnar cells rest. (*d*) The **goblet or beaker cells**.

6. The **mucosa**. (*a*) The **reticulum of the lymphoid tissue**. (It will appear only where the lymph-corpuscles have been accidentally brushed out.) (*b*) The **transversely divided ends of the elastic fibers**. (They appear as a pink mosaic.) (*c*) **Capillaries**. (They may frequently be traced for a considerable distance in their tortuous course.)

7. The cartilage plates. (*a*) Several cells in a single cavity. (*b*) The intracellular network.

8. The mucous glands. (*a*) That some of the cells are

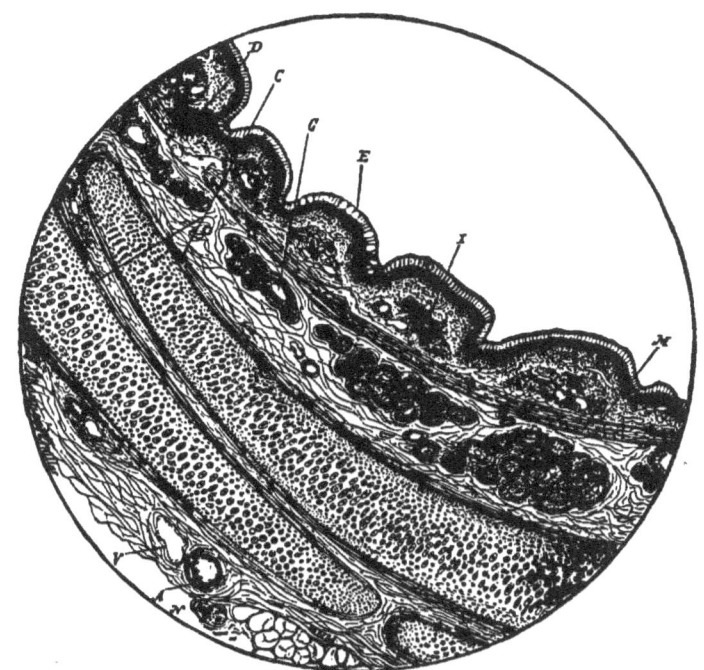

Fig. 85. Transverse Section of Part of the Wall of a Large Bronchus. Lung of Pig. Stained with Hæmatoxylin and Eosin (×60).

E. Epithelial lining. The line from the letter leads to a part of the lining containing large mucous cells or goblet cells.
I. The internal fibrous coat.
M. Muscular coat.
C. Cartilage plates of external fibrous coat.
A. Bronchial artery. The pulmonary artery is not included.
V. Bronchial vein.
N. Nerve-trunk.
G. Mucous glands.
D. Obliquely sectioned duct.

stained precisely like the other **mucous or goblet cells**, along the surface of the membrane. (*b*) If possible, a **gland-tube leading up to the lumen** of the bronchus. (An ampulliform dilatation is shown in the upper part of the drawing.)

THE PULMONARY BLOOD-VESSELS

The prominent accompaniments of the bronchus, at the root of the lung, are the pulmonary artery (carrying venous blood) and the pulmonary veins.

The pulmonary artery enters the lung with the bronchus, following its ramifications, to end in capillary plexuses in the walls of the sac-like dilatations, which are in connection with the ultimate bronchial tubes. The blood is then collected in venules, which unite to form the pulmonary veins. The latter pursue an independent course in their exit, not accompanying the bronchial tubes until the large bronchial tubes have been reached.

The bronchial artery (nutrient) enters with the bronchus, supplying its walls and the connective tissue framework of the lung.

A considerable amount of connective tissue accompanies and supports the structures which enter the lung, and is eventually in connection with the fibrous framework of the organ.

The lung will, therefore, be seen to differ from organs generally, in that it contains *two distinct vascular supplies*, viz.: (1) The *pulmonary* (of venous blood), entering for the purpose of its own oxygenation; (2) The *bronchial* (arterial), which corresponds to the usual nutrient blood-supply of organs.

THE PLEURA

The lung is completely enveloped in a membrane composed externally of endothelium, while the visceral portion is made up of interlacing fibrous and elastic tissue. The deep or visceral layer of the pleura sends prolongations in the form of septa into the substance of the lung, dividing it into rounded polyhedral compartments or lobules. The interlobular septa have usually become prominent in the human adult from deposits of inhaled carbon in their lymph-channels.

THE PULMONARY ALVEOLI

The lung is constantly employed in maintaining the integrity of the blood. This is accomplished by the exposure of the latter to a continual supply of atmospheric air. The air is introduced into little *sacs* (termed *air-vesicles* or *alveoli*), in the walls of which the blood is distributed in a *capillary plexus*. The air does not reach the capillaries themselves, inasmuch as they are covered with a layer of flat cells. These cells, constituting the *parenchyma* of the lung, have the power, on the one hand, of selecting such material from the air as may be required, passing it on to the blood in the capillaries; and, on the other, of removing effete material from the

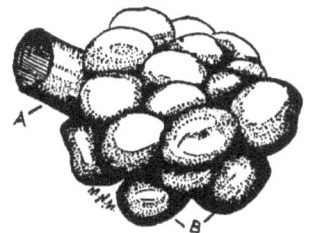

Fig. 86. Diagram of an Ultimate Pulmonary Lobule.
A. A terminal bronchiole.
B. The air-sacs or alveoli.

blood, transferring it to the atmospheric contents of the air-sacs for exhalation.

The air-sacs or alveoli are not unlike minute bladders. Their diameter about equals that of a terminal bronchus; viz., from one-fourth to one-eighth of a millimeter. A group of these alveoli are associated in the manner shown in Fig. 86, their contiguous walls

Fig. 87. Diagram Showing an Ultimate Pulmonary Lobule in Longitudinal Section, Showing the Manner in Which the Alveoli are Associated in Connection With a Terminal Bronchiole.
A. Terminal bronchiole entering.
B. The infundibulum.
C, C, C. Alveoli.

fusing and all opening into a common cavity, the *infundibulum*. The whole is in connection with a terminal bronchiole (*vide* Fig. 87). A *primary lobule* having been thus constructed, several are associated and united to a slightly larger bronchial twig, and there results one of the polyhedral lobules, previously mentioned as visible, especially on the surface of the lung. By a repetition of such elements the lung is constructed.

The wall of a pulmonary alveolus or air-sac is composed of connective tissue, supporting the capillary network, with a considerable amount of elastic tissue. The whole, as we have said, is lined with a single layer of flat, pavement epithelium. The capillary plexus, when filled with blood, affords the most prominent feature of the wall; but when the vessels have been emptied of their contents, they become very insignificant under the micro-

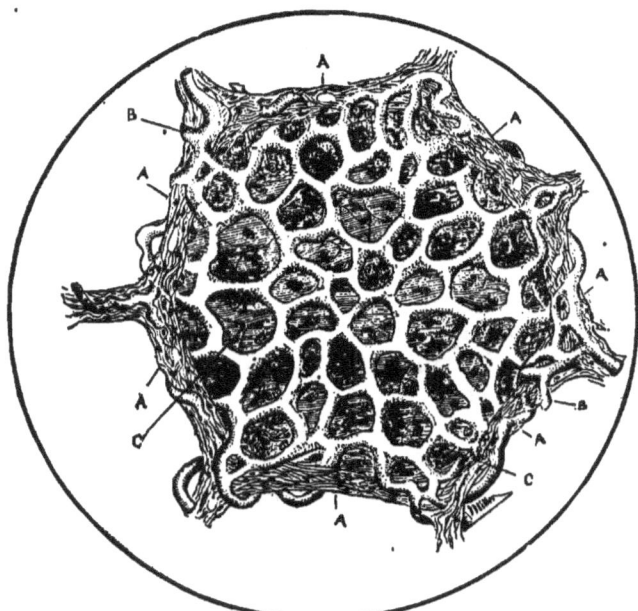

FIG. 88. TRANSVERSE SECTION OF A SINGLE PULMONARY ALVEOLUS. CAPILLARIES INJECTED. STAINED WITH HÆMATOLYLIN AND EOSIN (×400).

A, A, A. Walls of the alveolus.
B, B. Injected capillaries.
C, C. Pavement cells lining the alveolus. These cells cover the capillaries, but do not so appear in the drawing, as the latter are filled with an opaque injection. The observer is supposed to be above the sectioned alveolus, viewing the cup-shaped cavity.

scope, and the fibro-elastic tissue becomes more apparent. You will have observed that, aside from the vascular supply, the histology of an alveolar wall resembles very closely that of a terminal bronchiole, and when the vessels are all empty it is frequently difficult to differentiate them in the mounted section.

Fig. 88 shows a single alveolus, the vessels of which have been injected with a solution of colored gelatin. The alveolus has been divided through the middle, and shows as a cup-shaped cavity. The fibrous marginal walls are indicated, with their tortuous capillaries. The epithelial cells lining the bottom are obscured by the opaque capillaries, and shown only between the loops. It is probable that these cells cover the plexus completely, as they line the alveoli.

We now encounter an obstacle which will frequently be met in our study of organs. It consists of the difficulty in recognizing in sections the *plan of structure* which we have learned is peculiar to the organ under consideration. For example: A lung has been compared to a tree. The bronchi are the representatives of the branches, and the air-sacs of the fruit. Well, we make a section from human lung—it matters little as to the direction—with every possible care, and the image in the field of the microscope resembles a fragment of ragged lace more nearly than anything else! The arrangement of the tubes and alveoli of the lung has been determined by filling the cavities with melted wax, which, when cold, and the tissue destroyed by acid, gives a perfect mould of the organ. A *section* gives us but a single plane, and this fact *must be always borne in mind*.

PRACTICAL DEMONSTRATION

With a very sharp razor, cut centimeter cubes from pig's lung. Select portions free from large bronchi, with the pleura on one side at least, and harden with strong alcohol. Human lung, as fresh as possible, may be treated in the same manner. The epithelium of the alveoli shows best in young lung. Lung must be made very hard, or thin sections cannot be cut. If the ordinary ninety-five per cent. alcohol does not harden sufficiently, the process may be completed by transferring the tissue for twenty-four hours to absolute alcohol. The celloidin process is well adapted to this structure.

Stain the sections with borax-carmine, or hæmatoxylin and eosin. Mount in balsam.

SECTION OF LUNG OF PIG (*Vide* Fig. 89)
OBSERVE:
(L.)

1. The large scalloped openings, A, A, **transversely divided infundibula.**

2. The divided alveoli, B, B, so **sectioned as to** cut off both

bottom and top, and **show no epithelial lining excepting at inner edge of periphery.**

3. The alveoli, C, C, divided so as to show a **cup-shaped** bottom or top. (The minute granules are the nuclei of the lining cells.)

4. The alveoli, D, D, so cut as to leave most of bottom or top, **showing an opening in the center** where the sac has been sliced off.

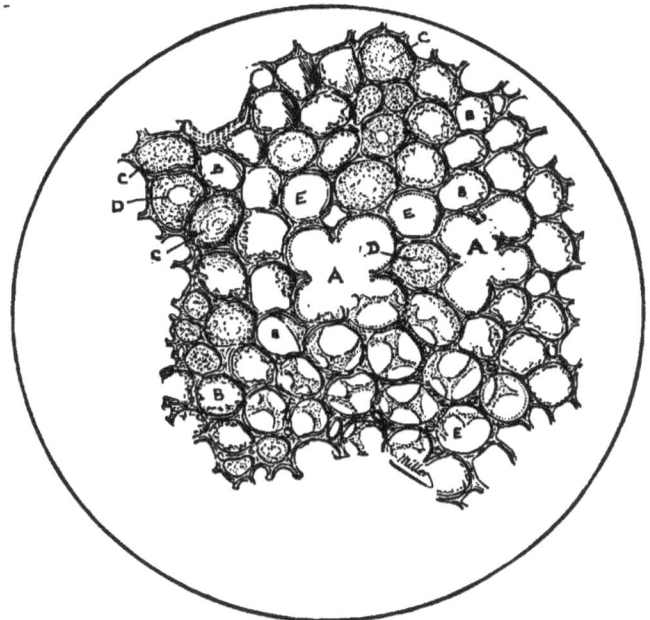

Fig. 89. Section of Lung of Pig. Stained with Hæmatoxylin and Eosin (×60).

A, A. Infundibula in T. S.
B, B, B. Alveoli; so sectioned as to show the outline only.
C, C, C. Alveoli; so sectioned as to present cup-shaped cavities.
D, D, D. Alveoli; so sectioned as to divide the top (or bottom).
E, E. Terminal bronchioles in T. S.

5. Openings, E, E, which are about the same size and bear a general resemblance to those of Obs. 2. (Note that their internal edges are smooth and not ragged. They are **terminal bronchioles.** No larger bronchial tubes have been included in the section.)

HUMAN LUNG—SECTION SHOWING A SINGLE ALVEOLUS (Fig. 90)

OBSERVE:

(L.)

1. The **outline of alveolus**. (The alveoli in human lung will show much distortion, as the tissue cannot be secured in perfect condition.)

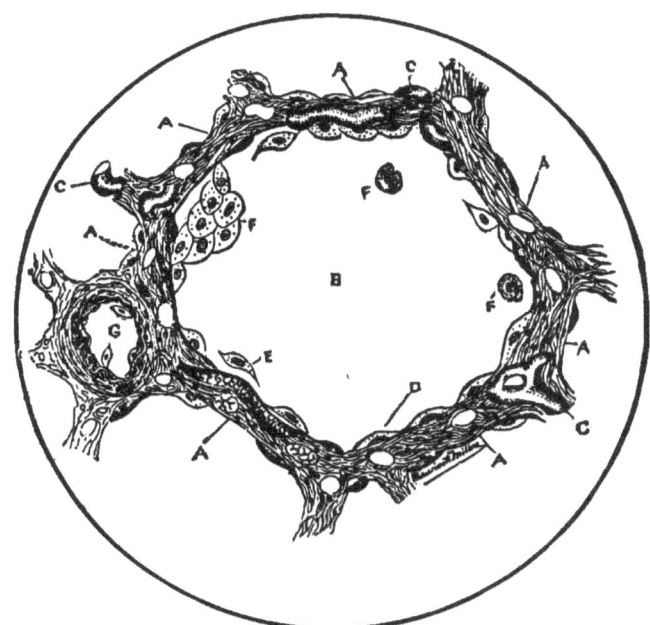

FIG. 90. TRANSVERSE SECTION OF A SINGLE PULMONARY ALVEOLUS. STAINED WITH HÆMATOXYLIN (× 400).

A, A, A. Walls of alveolus.
B. Lumen.
C, C, C. Capillaries variously sectioned in their tortuous course.
D. Pavement epithelial cells intact.
E. Detached pavement cell.
F. Detached cluster of pavement cells.
F'. Granular lining cells.
G. Pulmonary artery.

(H.)

2. The **fibrous wall**, A, A.

3. The **lumen**, B. (The bottom or top has been cut off in making the section.)

4. The **tortuous capillaries**, C, C, in the fibrous wall.

5 The **lining epithelial cells.** (*a*) Those **remaining attached** to the edges of the wall, D. (*b*) **Detached** cells, E. (*c*) Groups **partly detached,** F, F.

6. The divided **pulmonary artery,** G. (A medium-sized bronchial tube existed in the section immediately to the left of the artery.)

FŒTAL LUNG

Harden the lung of a fœtus, preferably human, in Müller's or Orth's fluid. After washing, finish hardening in alcohol; imbed in celloidin; cut; stain with hæmatoxylin and eosin; mount in balsam.

Observe the polyhedral form of the epithelial cells lining the alveoli. A few such polyhedral cells can be demonstrated in the alveoli of the adult lung with silver staining, and they correspond to the polyhedral cells of the terminal bronchioles. It appears that the plate-like cells lining the alveoli of the adult lung are derived from the polyhedral cells of the fœtal lung, which become flattened from the distension which they undergo when the alveoli are inflated.

THE TEETH

A human tooth is a calcareous structure of extreme hardness, and is divided into an exposed *crown*, a constricted *neck*, and one or more concealed *roots*—the latter being inserted into an alveolus, by means of which the whole is very firmly connected with the maxillary bone.

The central portion presents an elongated cavity (*pulp-chamber*) containing vascular, nervous, and connective tissue elements—the *pulp*.

The pulp-cavity is surrounded by the *dentine*, which constitutes the major portion of the tooth

The crown portion of the dentine is provided with a covering of *enamel*, while the root is invested with an osseous *cementum*, or *crusta petrosa*.

A thin membrane, 1μ or less in thickness—called the *membrane of Nasmyth* or the *cuticula*—covers the enamel in early life, while the cementum receives a periosteal investiture. The vascular and nervous elements of the pulp obtain admission to the pulp-cavity by a perforation or foramen at the apex of the root.

The Pulp.—The ground-substance, or stroma of the pulp, is a form of primitive connective tissue, gelatinous rather than markedly fibrous. It contains elongated capillary loops, multipolar cells, and medullated and non-medullated terminal nerve-fibrils.

Surrounding the pulp mass, and next to the dentinal wall of the chamber, we find a single layer of elongated cells—*odontoblasts*. These are probably in communication, by means of processes or prolongations, with fibrous elements of the pulp.

Dentine.—The dentinal stroma or matrix is made of fibrous tissue containing calcium salts, and is, next to the enamel, the hardest tissue of the body. The matrix is pierced with the *dentinal canals* (extremely minute channels, only 5μ in diameter), which radiate from their beginning, next the pulp-chamber, toward the outer portion of the dentine. These canals branch and anastomose, and are lined with an exceedingly thin *dentinal sheath*.

From the outer extremity of the odontoblasts of the pulp numerous prolongations are sent which are probably continued within the dentinal canals as the *dentinal fibers*. The dentinal canals terminate exteriorly, by very fine lumina, in a system of irregu-

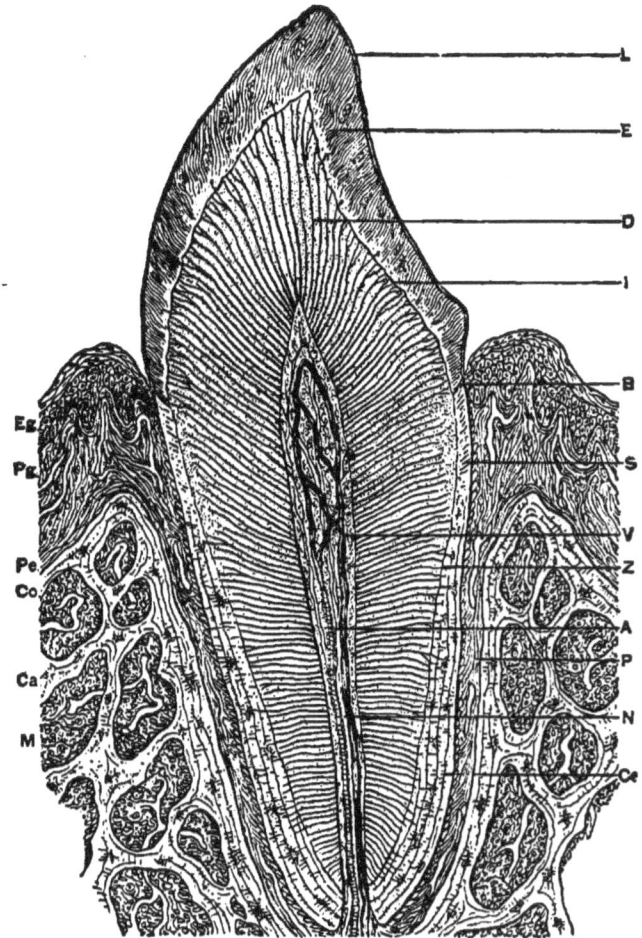

FIG. 91. DIAGRAM OF THE STRUCTURE AND IMPLANTATION OF A NORMAL INCISOR TOOTH. (BÖDECKER.)

    L. Cuticle of enamel, Nasmyth's membrane.
    E. Enamel.
    D. Dentine with uniformly distributed canaliculi.
    I. Interzonal layer between enamel and dentine.
    B. Border-line between enamel and cementum of neck.
    S. Cementum of neck.
    Ce. Cementum of root.
    Z. Interzonal layer between dentine and cementum.
    P. Pericementum.
    A. Arteriole of pulp, branching into capillaries.
    V. Vein of pulp taking up capillaries.
    N. Medullated nerve-fibers of pulp.
Eg. Stratified epithelium of gum.   Pg. Papillary layer of gum.
Pe. Periosteum.   Co. Cortical bone of alveolus or socket.
Ca. Cancellous bone-tissue of alveolus.   M. Medullary spaces of cancellous bone.

larly formed openings, *interglobular spaces*, which are channeled in the outer part of the dentine. The dentinal terminal fibers are in connection with branched cells which occupy the interglobular spaces.

The Enamel.—The part of the dentine above the neck of the tooth is protected by a covering of *enamel*. The enamel consists of *prisms*, 4 μ in diameter, united into bundles by a little cement substance, which pass in a direction nearly at a right angle to the surface of the dentine. They are of extreme density, contain little besides inorganic material, and in a vertical section the whole is traversed by parallel striæ, not unlike the markings indicating tree-growth—the *lines of Retzius.*

Cementum.—The fang portion of the dentine is invested with a thin layer of true bone, containing *lacunæ* and canaliculi, but no Haversian canals. The cementum is provided with *pericementum* (periosteum), which forms the bond of union between the teeth and the process of the maxillary bone. The bone-corpuscles are in connection, through the canaliculi, with the cells in the interglobular spaces of the dentine. It will be seen that the connective tissue elements, at least of the pulp, are in eventual histological connection with the bone-corpuscles of the cementum.

PRACTICAL DEMONSTRATION

The illustrations given in text-books have been drawn from dried teeth, ground down to the requisite thinness by means of corundum or emery wheels. This is a very tedious process, and is impracticable with the student. If such specimens are desired, it will be advisable to purchase them already mounted. They only give the skeleton of the organ, all the soft tissues being destroyed by the drying and grinding.

While dry specimens exhibit the plan of a tooth, the soft tissues must be studied in sections made after the inorganic constituents have been removed. Teeth immediately after extraction are to be treated in the same manner as described for bone. A one-sixth per cent. chromic-acid solution, to which five drops of nitric or hydrochloric acid have been added, may be used. Let the quantity of liquid be liberal, and from time to time, say every three days, add a few drops of the nitric acid. The decalcification should proceed slowly, and may be complete in from two to three or four weeks. The earthy matters will first be dissolved from the surface. Watch the action carefully, ascertaining the progress of decalcification by pricking a fang with a needle. If the acid be too strong, and the action too rapid, the whole may be destroyed. When the decalcification is complete, a needle may be easily passed through the tooth, and sections may be made with the razor or knife, with or without a microtome. The form will be preserved except as regards the enamel; this will be entirely dissolved. The enamel prisms may be demonstrated by treating broken fragments with dilute acid for a short time only.

Sections should be stained with carmine and picric acid and mounted in glycerin. For the study of the development of teeth, fœtal jaws may be treated as just described; and, when properly decalcified and hardened, should be infiltrated with celloidin, sectioned, and stained.

TRANSVERSE SECTION OF FANG OF HUMAN DECIDUOUS CANINE TOOTH—DECALCIFIED (Fig. 92)

OBSERVE:

(L.)

1. **Division into pulp, dentine, cementum, and pericementum.**

2. **Line of junction of pulp and dentine.** (If the elements of the pulp are intact, note the **layer of** deeply stained **odontoblasts** next the dentine.)

3. **External limit of dentine.** (Note here the deeply stained granular **line of Purkinje.** This is the location of the interglobular spaces. The deep color is due to the staining of the contents of their cells.)

4. The **striae of the dentine.** (The dentinal canals and stained contents.)

5. The **laminated cementum.** (The yellowish pink dots on the lacunæ.)

(H.)

6. **Elements of the pulp.** (*a*) The layer of **odontoblasts.** (Note their **internal processes** connecting with other cells of the pulp; and the **external processes** passing into the dentinal canals.) (*b*) The **sparsely fibrillated character of the pulptissue.** (*c*) Sections of **vascular loops.** (The nerve-elements may be demonstrated, particularly if the section be made near the apex of the root, where the fibers are medullated. The terminal fibrillæ are non-medullated.)

7. **Dentinal elements.** (*a*) The **dentinal canals.** (*b*) The **dentinal sheath.** (Better demonstrated in transverse sections.) (*c*) **Dentinal fibers.** (In transverse sections the canals are well shown lined with a membrane of extraordinary tenuity, with the fiber appearing as a central dot.) (*d*) **Fine dentinal fibers** near the outer limit. (*e*) **Interglobular spaces.** (An occasional cell may be made out in the larger spaces. They were formerly sup-

posed to contain a gelatinous material only. Note the connection between these spaces and the termini of the dentinal fibers.)

8. The **cementum**. (*a*) The **lacunae**. (*b*) **Bone-corpuscles** in the last. (The canaliculi are not well demonstrated here, as the tissue is very translucent and feebly stained. These minute canals are better indicated in dried bone.)

9. **The pericementum.** (Note its dense fibrillar meshwork.)

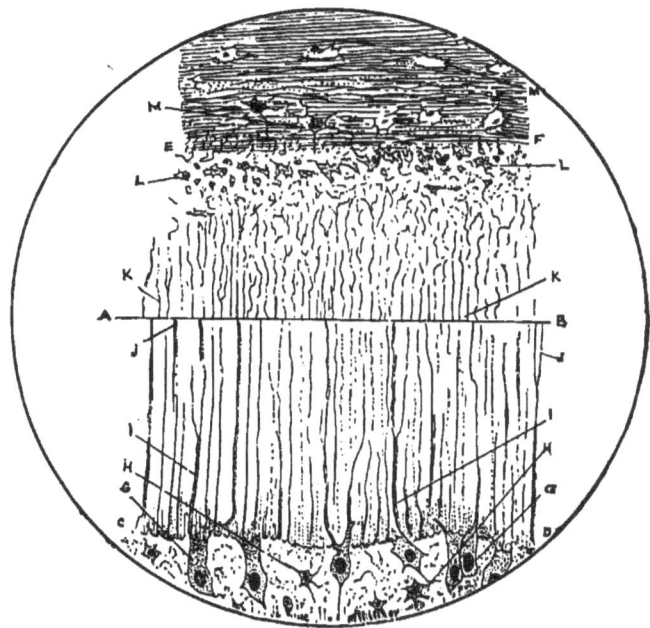

FIG. 92. TRANSVERSE SECTION OF FANG OF A HUMAN DECIDUOUS CANINE TOOTH, DECALCIFIED WITH CHROMIC AND NITRIC ACIDS AND STAINED WITH PICROCARMINE ($\times$ 400).

A, B. Line through the dentine indicating the point at which the edges have been made to join after the omission of an intervening portion. This was necessary in order that the different layers might be shown in a single drawing.

C, D. Junction line between the pulp and dentine.

E, F. Junction line between dentine and cementum.

G, G. Odontoblasts of the pulp.

H, H. Stellate connective tissue cells of the pulp.

I, I. Dentinal processes of odontoblasts.

J, J. Dentinal fibers.

K, K. Terminal branching dentinal fibers.

L, L. Interglobular spaces of dentine.

M, M. Lacunæ of the cementum. The drawing does not show the periosteal investiture of the crusta.

GLANDS

A gland is an organ—frequently subsidiary to and located within other organs—whose cells manufacture from the blood products to be utilized in the performance of some of the functions of the body, or waste products which are to be excreted.

Simple glands are tubes or cavities, with connective tissue walls lined with epithelial cells, which are usually placed upon a basement membrane. Around, and in close proximity to the lining, is spread a plexus of blood-capillaries. In *compound glands*, the simple glands (acini or alveoli) are enclosed by connective tissue in groups called *lobules*, and larger groups called *lobes*. The same connective tissue is continued to form a *capsule* over the outside.

The essential parts of a gland are, therefore:
1. A *duct*, or efferent conduit for the secretion.
2. *Parenchyma*, or cells engaged in secretion.
3. A *blood-vascular supply*.

TUBULAR GLANDS

The simplest gland-structure is offered in the form of a tube. Glands are, frequently, little more than tubular depressions in mucous surfaces. Examples are found in the uterus, and small and large intestines.

COILED TUBULAR GLANDS

Tubular glands are often greatly elongated, with the blind extremity coiled. This variation presents the simplest differentiation between the part of the tube which is secretory, and the duct, or drainage part. With this change in function of the different extremities of the tube will occur a change of epithelium. The cells belonging to the duct-end will usually retain the columnar form; while the actively secreting elements will become enlarged, more nearly filling the tube, and assume a polyhedral form from pressure.

Examples have already been seen in the sweat-glands of the skin.

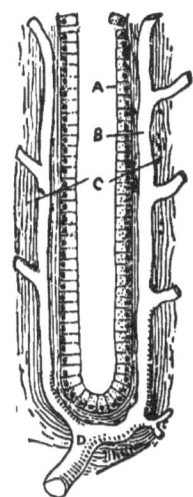

FIG. 93. DIAGRAM. SIMPLE TUBULAR GLAND.

A. Lining cells—parenchyma.
B. Capillary plexus, supplying the parenchyma.
C. Connective tissue supporting capillaries.
D. Arterial supply.

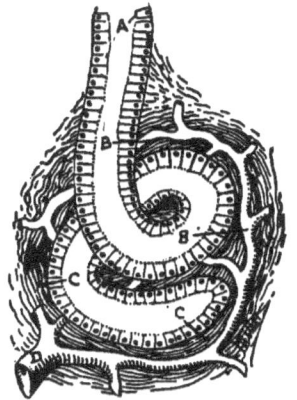

FIG. 94. DIAGRAM. COILED TUBULAR GLAND.
Same references as Fig. 93.

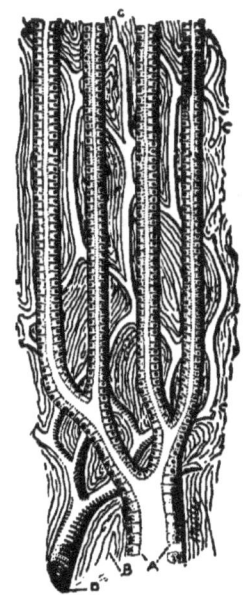

Fig. 95. Diagram. Branched Tubular Gland.
References same as Fig. 93.

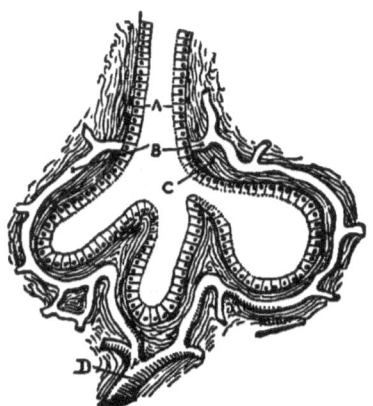

Fig. 96. Diagram. Illustrating the Plan of Acinous Glands.
References same as Fig. 93.

BRANCHED TUBULAR GLANDS

With the branching of the duct-portions of gland-tubules, there usually occurs a dilatation of the extremities into acini or alveoli, although pure examples of branched tubular glands are afforded in the gastric and uterine glands.

The most nearly typical branching of gland-like tubules is afforded by the tubuli uriniferi of the kidney, or the system of tubes in the testicle. The tubules here present other features peculiar to them, which will be referred to under the proper head.

ACINOUS GLANDS

The dilatation of branching tubules, referred to under the previous heading, results in the formation of acinous glands. They are formed by the subdivision of a main tube or duct, with repeated branching of the secondary tubules. Collections of terminal branches often result in globular masses, which are more or less perfectly isolated from one another by connective tissue. In this way *compound acini* are produced, such as the pancreas, the salivary, mammary, and buccal glands.

The large compound acinous glands are also called *compound racemose glands*. Simple acinous glands do not occur in human tissues.

THE PAROTID GLAND

The *parotid, submaxillary, sublingual,* and *buccal salivary glands* are typical glandular structures, with individual peculiarities only in respect to the cell-elements; these vary according to the nature of the secretion formed in each.

The parotid is a *compound acinous gland*, leading from which is a principal duct—lined with tall columnar cells—which collects the fluid saliva from the different divisions of the organ.

As the duct penetrates the gland it branches freely, the lumina becoming smaller and the cells shorter as the deeper parts are approached.

Each terminal duct is in connection with several acini. The connective tissue adventitia of the duct becomes the thin wall of the acinus, and the lining cells broaden, frequently become poly-

144 STUDENTS HISTOLOGY

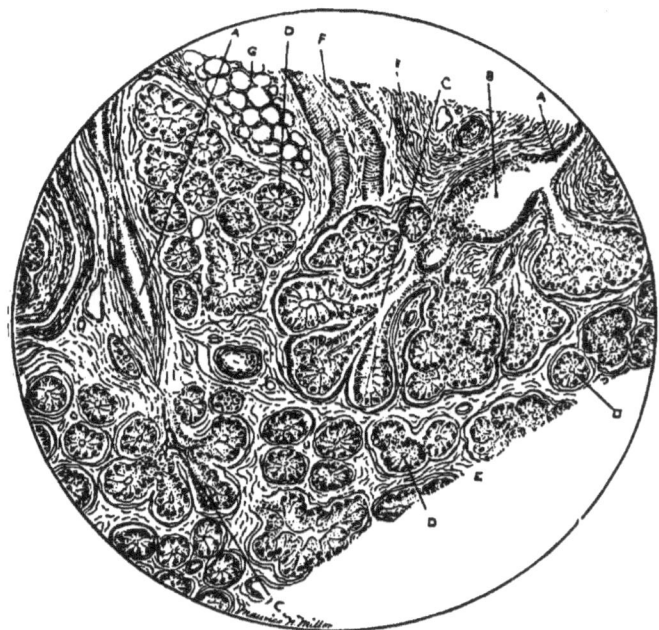

FIG. 97. SECTION OF A SMALL PORTION OF THE PAROTID GLAND. STAINED WITH HÆMATOXYLIN AND EOSIN ($\times$ 250).

- A. Narrowing of the duct from a small lobule, before entering a larger duct.
- B. Dilatation of a duct after leaving a small lobule.
- C. Primary lobules, in nearly L. S.
- D. Acini in T. S., showing the minute lumen.
- E. Connective tissue supporting the gland.
- F. Striated muscular fiber adjacent to the gland.
- G. Adipose tissue in the loose areolar tissue.

hedral, and are bluntly pointed. The cells so nearly fill the acini as to leave a small and not easily recognized lumen.

The gland is richly supplied with blood-vessels.

THE SUBMAXILLARY GLAND

The submaxillary is presented as an example of a typical *mixed gland*. The general arrangement is not unlike that of the other salivary glands. Its peculiarity appears in the parenchyma, and will be noticed later.

Pure mucous glands are found in the submucosa of the mouth, tongue, fauces, trachea, and the larger bronchi.

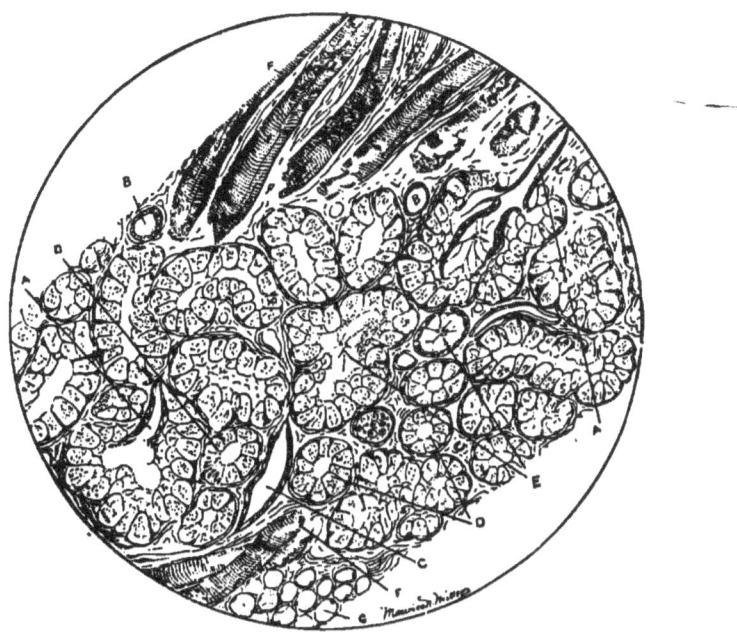

Fig. 98. Section of Part of the Submaxillary Gland (× 250).

- A. Narrow duct from terminal lobules.
- B. Small duct in T. S.
- C. Small duct in oblique section.
- D. Transversely divided acini, showing large lumen.
- E. Mucus remaining in the lumina.
- F. Striated muscular fibers.
- G. Adipose tissue.

THE PANCREAS

The histology of the pancreas is, in general, that of a true *serous gland*, like the parotid. It has been called by physiologists the abdominal salivary gland.

The lobules are more tubular and less regular in size and form; and the lumen of the acini is much less easy of demonstration, in an ordinary hardened section, than the same in the parotid.

The branches of the pancreatic duct are provided with a very thick adventitia, are lined with short columnar cells, and seldom present the dilatation which generally occurs in a serous gland on entering the lobule.

J

PAROTID AND SUBMAXILLARY GLANDS, AND THE PANCREAS

Practical Demonstration

The tissue must be fresh, divided in small pieces—not larger than a centimeter cube—and hardened by placing in ninety-five per cent. alcohol for twelve hours, after which fresh spirit should be substituted. If, after the lapse of another twelve hours, the tissue should not be sufficiently firm, it should be placed in a small quantity of absolute alcohol for three hours. Sections should

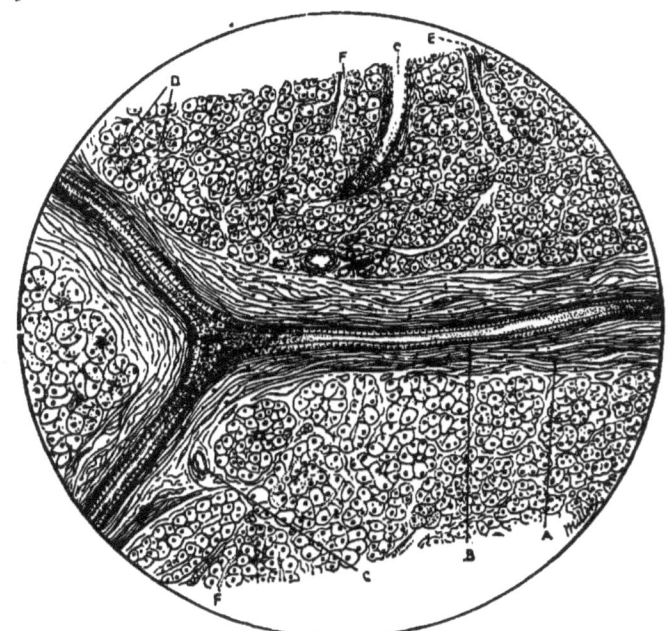

FIG. 99. SECTION FROM THE PANCREAS.

A. Wall of a large duct.
B. The somewhat cubical lining cells.
C. Arteries.
D. Lumen of the aciui, T. S.
E. Terminal duct leaving a lobule.
F. Acini in L. S.

be made immediately after hardening—as more prolonged action of the strong spirit will cause the tissue to contract.

Sections may be cut with or without a simple microtome—the desideratum being *thin rather than large cuts*.

Stain lightly with hæmatoxylin and deeply with eosin.

After sections of hardened tissue have been examined, the glandular parenchyma may be profitably studied in teasings from tissue which has been in Müller twenty-four hours. Wash the teasings on the slide with a liberal supply of water, removing the same from time to time with blotting-paper. Add a drop of hæmatoxylin solution; and, after washing this away, add a drop of glycerin, and cover. This method is very generally useful for teased or scraped fragments of glandular structures.

OBSERVE: (Figs. 97, 98 and 99)

(L.)

1. The **connective tissue.** (Most abundant in the parotid gland, and least so in the pancreas.)

2. The **ducts.** (Note the **flattening of the lining columnar cells,** as the ducts approach the acini, until **mere scales** result. Also the thick connective tissue adventitia, especially demonstrable in the pancreas.)

3. The **lobules.** (These are formed by several acini, and are most typical in the parotid. It must be remembered that only one plane is visible, and that there is little perspective.)

4. The **acini.** (**Note the lumina—large** in the **submaxillary, less so in the parotid,** and **least,** and often difficult to make out, **in the pancreas.**

5. The **blood-vessels, muscular and adipose tissue.** (The two latter are demonstrable only in the salivary glands, and do not belong properly to the gland itself. The capsule of the pancreas, in common with such structures in general, contains adipose tissue. The abundant interacinous capillary plexuses of the pancreas require the high-power for satisfactory demonstration.)

(H.)

6. The **parenchyma.** (*a*) The small but distinct **shortened columnar cells** of the acini **of the parotid.** (Observe that they are frequently so formed that the convexity of one cell fits into the concavity of its neighbor. Where seen in transverse section, the outline is a polygon. Note especially the **change in** the parenchymatous elements as the **terminal duct** merges into an acinus.)

(*b*) The **large, swollen cells of the mucous acini—submaxillary.** (Observe the comparative **clearness** of the cells. They contain a very delicate **reticulum,** and their **nuclei** are often **obscured** and frequently seen to be **placed at the junction of the cells.**)

(*c*) The **rounded,** often polyhedral **cells of the pancreas.**

(They resemble the parotid elements, although smaller and less granular. The acini are more tubular than in the parotid gland. Even with the low-power one may distinguish small, rounded areas called the bodies of Langerhans. They are probably groups of immature acini.)

(d) In the smaller ducts of the parotid and submaxillary glands, notice that the lining epithelial cells have vertical striations at the outer border.

The sublingual gland in man is an almost purely **mucous** gland. The parotid is purely **serous**, and has a secretion of that character. It is reckoned as a **true salivary gland**. The submaxillary, having elements of both kinds, is called **mixed**. In the serous glands, when the cells are filled with secretion, they appear large, with fine granules. After discharge of the secretion they are smaller, dark, and granular. The cells of the mucous glands may be found large and clear, or after discharge of the mucus, also smaller, dark, and granular. At the borders of the acini of the mucous glands there may be seen crescent-shaped groups of granular cells, **the demilunes of Heidenhain**.

THE THYROID GLAND

The thyroid gland is a compound tubular gland. The tubular acini are 40 to 120 μ in diameter. Loose connective tissue unites them into lobules and lobes, and forms a covering for the whole. The acini are closed cavities. They are lined by low columnar or cubical epithelial cells. The cells are limited by a basement membrane. The acini are usually filled with homogenous, yellow, translucent material, called the *colloid* substance, which is formed by the epithelial cells. The blood supply is abundant, and a rich capillary network surrounds the acini.

The lymphatic network is also profuse, and the characteristic colloid substance may be found within the lymphatics. Nerve-fibers are not numerous. They are mostly non-medullated, derived from the sympathetic. Although the thyroid gland of the adult has no duct, in the embryo it has one—the *thyro-glossal duct*, which has an opening corresponding to the foramen cæcum, near the base of the tongue. Later on, this duct becomes obliterated for the most part.

THE MOUTH AND PHARYNX

The *mouth* cavity is lined by a mucous membrane, consisting of a stratified squamous epithelium and a connective tissue layer below it. The connective tissue layer possesses minute papillæ resembling those of the skin. Numerous mucous glands open into the oral cavity, as well as the salivary glands.

The *tongue*, which is composed chiefly of striated muscle, exhibits on its upper surface very large papillæ. These papillæ are of three sorts: (1) *Filiform*, or conical; (2) *fungiform*, or those having a constricted base, and (3) *circumvallate*, which are only eight or ten in number, arranged like an inverted V, at the back of the tongue. The last variety are large, fungiform, and surrounded by a depression, outside of which is a wall-like elevation. *Taste-buds* are flask-shaped collections of epithelial cells, specialized for the perception of taste, and supposed to be connected with nerve-fibers. They occur at the sides of the circumvallate papillæ and elsewhere on the tongue. They are most easily studied in sections of the papillæ foliatæ of the rabbit, which are symmetrical, oval areas, marked with parallel ridges, at the back of the tongue. There is an abundance of *lymphoid tissue* at the back of the tongue, either diffused or occurring as distinct spherical lymph-follicles.

The *tonsils* are large collections of lymphoid tissue, containing numerous denser spherical masses, *lymph-follicles*. Stratified squamous epithelium covers the free surface of the tonsils and lines certain blind depressions into the tonsils (crypts). The epithelium is more or less infiltrated with lymphoid cells, which enter it from the underlying tissue. The attached surface of the tonsil is covered by a connective tissue capsule, which forms an adventitia.

The so-called salivary corpuscles are lymphoid cells which have escaped and become mixed with the saliva.

The structure of the *pharynx* is nearly like that of the mouth. Only the lower division, however, is lined with stratified squamous epithelium. The portion above the level of the soft palate is covered with stratified columnar epithelium, which is also ciliated, indicating its connection with the respiratory tract. The mucous membrane of the pharynx also contains mucous glands and lymphoid tissue. A mass of lymphoid tissue occupying a position in the upper part between the Eustachian tubes is called the *pharyngeal tonsil* of Luschka.

THE ŒSOPHAGUS

Beginning with the œsophagus, we encounter an arrangement of layers of muscle and fibrous tissue, constituting the walls of a tube, lined with mucous membrane, which continues throughout the rest of the alimentary canal. The part of the tube below the œsophagus possesses in addition a serous covering, derived from the peritoneum, on the outside.

The walls of the œsophagus are made of the following layers from within out: Mucous, muscularis mucosæ, submucous, muscular, and fibrous.

The *mucous membrane* is covered with stratified squamous epithelium, resting on a layer of connective tissue, which presents minute papillæ.

The *muscularis mucosæ* consists of a small amount of unstriated muscle, lying below the mucous membrane. The fibers run longitudinally.

The *submucous* layer is composed of loose connective tissue, which accommodates itself to the contractions of the muscular portion, and permits the mucous membrane to be thrown into folds. The blood-vessels, lymphatics, and nerves are distributed through the submucous layer to the other structures. Mucous glands are found in the submucous tissue. Their contents reach the surface by means of ducts.

The *muscular* coat consists of an inner layer, fibers of which run in a circular direction about the œsophagus, and of an outer longitudinal layer. In the upper third of the œsophagus the muscle is striated, in the lower third it is unstriated, and in the middle third the two kinds are found mixed.

PRACTICAL DEMONSTRATIONS

The tongue, taste-buds, tonsils, and œsophagus should be studied in this connection. The taste-buds may easily be obtained in sections of the papillæ foliatæ of the rabbit's tongue. The tongue itself should be secured from one of the lower animals, and sections that pass through the papillæ should be sketched. The papillæ make beautiful objects in stained sections. The tonsils of a human subject should be used. The sections of the tonsil must be very thin. The œsophagus may be taken from one of the lower animals. These tissues may be hardened in alcohol, or, better, in Orth's mixture of Müller's fluid and formaldehyde. They may be cut, stained with hæmatoxylin and eosin, and mounted in the usual manner.

THE STOMACH AND INTESTINES

The stomach and intestines are lined with mucous membrane; *i.e.*, a membrane containing epithelial cells, usually of the columnar variety, which secrete mucus.

The gastric and intestinal walls are constructed as follows:
1. *The epithelial lining.*
2. *The mucosa.*
3. *The muscularis mucosæ.*
4. *The submucosa.*
5. *The muscular walls proper.*
6. *The fibrous or peritoneal investment.*

In descriptive anatomy, the first three of the above are included in the mucous coat.

The epithelium of the inner surface of that portion of the alimentary tract under consideration is of the columnar variety. Variations occur in the deeper layers, which will be referred to later on.

The mucosa, with its epithelial covering, is thrown into coarse folds, *rugæ* or valve-like reduplications, which greatly increase the extent of surface. It contains the principal glands and capillary blood-vessels. The epithelial lining is usually considered as a part of the mucosa.

The muscularis mucosæ is a thin layer of involuntary muscular fiber, which separates the mucosa from the submucosa. Some of the cells run in a longitudinal and others in a circular direction.

The submucosa, composed of loose areolar tissue, serves to connect the previous structures with the muscular coat proper, and contains the larger trunks from which the capillaries of the mucosa take their origin, or into which they empty. An intricate plexus of lymphatics is also situated here.

The muscular coat consists of strong bands, running chiefly in two directions, corresponding to an inner circular and an outer longitudinal layer. Near the cardiac end there is an imperfectly developed oblique layer. The muscle-plates are sustained by connective tissue.

A peritoneal investment covers the organs, except at such points as are occupied by the entrance and exit of blood-vessels and lymphatics through the mesenteries, with a few exceptions.

THE STOMACH

The mucosa everywhere contains microscopical depressions, the *gastric tubules* or *glands*. These are concerned in the production of the gastric juice.

The several layers of the stomach may be better understood by reference to the diagram (Fig. 100).

The gastric tubular glands are of two principal varieties; viz., 1, the *peptic glands*, found in the cardiac portion of the stomach; 2, the *pyloric glands*, which occupy the pyloric extremity of the

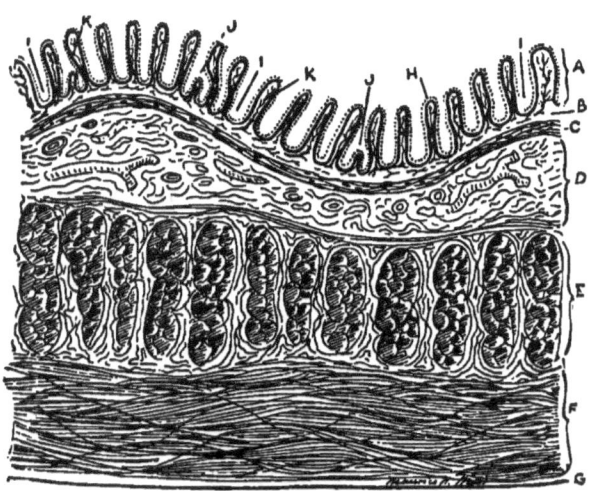

FIG. 100. DIAGRAM OF THE WALL OF THE STOMACH IN VERTICAL SECTION.

    A. Layer of gastric tubules.
    B. Vascular portion of mucosa.
    C. Muscularis mucosæ.
    D. Submucosa.
    E. Internal circular layer of muscular fiber.
    F. External oblique and longitudinal muscular layers.
    G. Peritoneum.
    I, I, I. Lumen of gastric tubules.
    J, J. Branching gastric tubules.
    K, K. Blood-vessels arising from lower portion of mucosa
        forming plexuses between the tubules.

organ. The mucous membrane, midway between the cardiac and pyloric portions, is occupied by tubules which partake of the character of both peptic and pyloric glands, so that no sharp boundary line exists.

The peptic or cardiac gland-tubes penetrate to the muscularis mucosæ. They pursue a somewhat wavy course, and at their lower or blind extremity are frequently bifid. They are lined at their commencement on the surface with translucent columnar epithelium, the cells being polygonal in transverse section. As the fundus or bottom of the tube is approached, the lining cells become granular, larger, and somewhat polyhedral. Next the wall of the tube large, granular, bulging cells are scattered irregularly. The epithelium occupies the major portion of the space in the tube, so that the lumen is very small.

A single bifid tube is represented in Fig. 101. The prominent

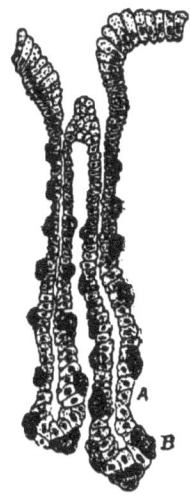

FIG. 101. VERTICAL SECTION OF A PEPTIC TUBULAR GLAND, FROM CARDIAC MUCOSA OF STOMACH. LARGELY DIAGRAMMATIC.

A. Central or chief cells.
B. Border or parietal cells.

distinguishing feature of the peptic or cardiac tubules is afforded by the large *border* or *parietal cells*. The cells next the lumina are called *central* or *chief cells*.

The pyloric gland-tubes pursue a course not greatly unlike that of the tubes just mentioned. They do not branch, however, until they have penetrated well down toward the muscularis mucosæ. Their distinguishing character is afforded by the epithelial lining. At the surface, the cells are columnar, with polygonal transection.

The deeper parts are lined with translucent cylinders. The lumina are larger than those of the peptic tubes.

The gastric gland-tubes are placed thickly side by side, their bases reaching the muscularis mucosæ. Between and beneath the tubes is a dense network of blood-capillaries.

The remainder of the stomach has little special interest for the histologist. The muscular portion of its walls consists of a thin

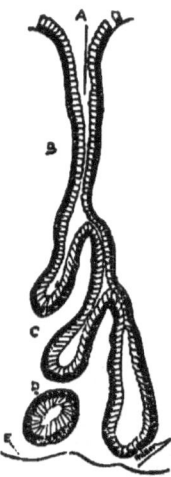

FIG. 102. VERTICAL SECTION OF TORTUOUS AND BRANCHING TUBULAR GLAND, FROM PYLORIC MUCOSA OF STOMACH. DIAGRAMMATIC.

    A. Lumen. This is often much widened.
    B. Duct portion of tubule.
    C. Branching glandular portion, or fundus.
    D. Transverse section of the fundus.
    E. Lower limit of mucosa.

internal circular layer, with oblique bundles interspersed, and a thin external longitudinal layer, both being of the involuntary variety. Between the two layers is found a plexus of non-medullated nerves, corresponding to the plexus of Auerbach, and another in the submucosa, corresponding to the plexus of Meissner of the intestines, but they are not usually demonstrable by ordinary methods or sections.

The blood-supply is received at the curvatures. Branches penetrate the muscular layers along the lines of omental attachment, as blood-vessels never penetrate the peritoneum.

The peritoneum is constructed mainly of fibrous tissue, with an external investment of endothelium.

THE STOMACH

PRACTICAL DEMONSTRATION

Inasmuch as the human stomach cannot often be obtained until decomposition has destroyed it for our work, we must secure the organ from some one of the lower animals. The stomach of the dog presents all the histological features of that of man, and can be obtained in good condition from an animal recently killed.

Harden small pieces in strong alcohol, and cut sections at a right angle to the surface and from different regions. Stain with hæmatoxylin and eosin, and mount in balsam.

VERTICAL SECTION OF WALL OF CENTRAL PORTION OF DOG'S STOMACH (Fig. 103)

OBSERVE:

(L.)

1. The division into: (*a*) **Surface epithelium** (free ends of gland-tubes). (*b*) **Mucosa.** (*c*) **Muscularis mucosae.** (*d*) **Sub-**

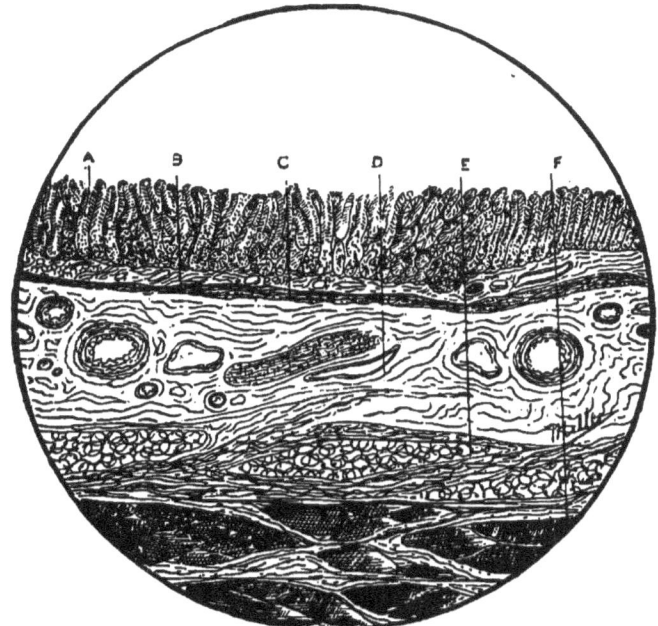

FIG. 103. VERTICAL SECTION OF WALL OF CENTRAL PORTION OF DOG'S STOMACH.

A. Internal surface, showing open mouths of the gastric tubules, lined with clear columnar cells.
B. Deepest portion of mucosa.   C. Muscularis mucosæ.
D. Submucosa.   E. Adipose tissue in last.
F. Bundles of muscular tissue (internal circular) ($\times$ 60).

mucosa. (*e*) **Muscular layers.** (Only a portion of the inner circular layer is shown. It has been divided transversely.) (**H.**)

2. The **epithelium of gland-tubes.** (The upper portion of the tubes will be cut obliquely in many places, as they have been inclined, and the epithelium will show as a beautiful mosaic of polygonal areas.) (*a*) The differentiation between **border** and **central cells.** (*b*) **Tubes cut transversely,** showing the lumina. (*c*) Indications of the **capillary plexuses between the tubes.**

3. The **mucosa.** (*a*) **Arterioles and venules** beneath the tubules. (*b*) Scattered **lymphoid cells** (round cells with one, two, or three nuclei).

4. **Lenticular glands,** masses of **adenoid or lymphoid tissue** at the bottom of the mucous membrane, chiefly near the pylorus.

5. The **muscularis mucosae.** (Note the elongated nuclei of the smooth muscle-cells.)

6. The **submucosa.** (*a*) **Arteries, veins,** etc., cut in various directions. (*b*) The **adipose tissue.** (Fat-crystals are frequently seen in the cells when freshly mounted.)

7. The **muscular bundles of the circular layer** with the **septa of connective tissue.** (Note particularly the various appearances presented by bundles of involuntary muscular fiber when cut in different planes.)

8. Groups of **ganglion cells** belonging to the plexuses of Auerbach or Meissner will occasionally be seen on very careful examination.

SMALL INTESTINE

The histology of the intestines, both large and small, is formed upon the general plan of that of the stomach. The same layers are presented: the *mucosa*, with its epithelial covering; the *muscularis mucosæ;* the *submucosa;* the *muscular* and *peritoneal* coats.

The mucosa of the small intestine is everywhere pierced by blind depressions; and the surface is studded with minute elevations or papillæ, between which are the *depressions which correspond to the tubules of the stomach*. The elevations are called *villi*, the depressions between the villi, *the crypts* of Lieberkühn.

The small intestine serves two important functions: 1. The *secretion of a fluid*, one of the digestive juices—the *succus entericus*. 2. The *absorption of food*, especially the fats or hydrocarbons.

We shall view the histology of this organ from a physiological standpoint, considering: (1) *Those structures concerned in the secretion of the succus entericus;* (2) *Those portions concerned in absorption of food.*

HISTOLOGY OF THOSE PARTS OF THE SMALL INTESTINE PARTICULARLY CONCERNED IN THE PRODUCTION OF THE SUCCUS ENTERICUS.

The diagram (Fig. 104) is intended to represent at A the thickness of the mucosa with its papillary elevations—the villi. The muscularis mucosæ B, from which the villi arise, separates the mucosa from the submucosa C. The horizontal line at the bottom of the diagram indicates the outer limit of C and the beginning of the circular muscular coat of the intestine. The *villi*, everywhere covered with columnar epithelium, often containing goblet-cells, are represented in the drawing as widely separated. The crypts of Lieberkühn, which are lined by columnar epithelium, open between the prominences. In the interior of each villus is a fine network of *blood-capillaries* (G, G). (In the specimen from which the sketch was made the blood-vessels had been injected with colored gelatin to make them prominent.) The cells on the borders of the crypts secrete certain fluid material from the blood circulating in the capillary plexuses, and pour it out into the crypts; the crypts becoming filled with the fluid, the latter overflows and passes into the lumen of the gut, to act in promoting digestion. This is one source of the succus entericus, and there is yet another.

Between the bases of some of the villi, tubes or ducts will be found which, piercing the muscularis mucosæ, reach the submucosa, where they branch, become convoluted, are lined with secreting cells, and are known as the *glands of Brunner*. They occur in the duodenum and are continuations of the pyloric glands of the stomach. These glands are surrounded by blood-capillaries, and the gland-cells secrete a fluid which is poured into the gut between the villi, when it becomes mingled with the secretion previously mentioned, and constitutes a part of the *succus entericus*.

We have, then, seen that the succus entericus is secreted *partly from the epithelial cells lining the crypts of Lieberkühn*, and *partly from the cells of Brunner's glands*.

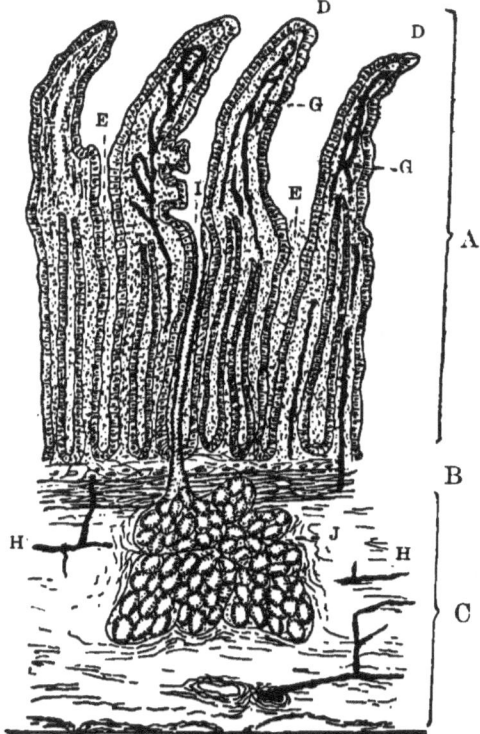

Fig. 104. Diagram showing Portions of Intestinal Mucous Membrane concerned in the Secretion of the Succus Entericus.

- A. The mucosa.
- B. Muscularis mucosæ.
- C. Submucosa.
- D, D, D. Villi.
- E. Crypts of Lieberkühn.
- G. Blood-plexuses of villi.
- H, H. Large vessels of submucosa, supplying the epithelium covering the villi.
- I. Neck of a gland of Brunner.
- J. Gland of Brunner in the submucosa.

THE REMAINING STRUCTURES OF THE INTESTINE CONCERNED MAINLY IN FOOD ABSORPTION

The diagram (Fig. 105) is intended to show the same layers as were indicated in the previous figure (Brunner's glands and the blood-vessels have been omitted in order to avoid confusion). The villi are represented much shortened.

In the center of each villus is the blind tube G, G, a part of the lymphatic system, and here called a *lacteal*. When, during digestion, the minute globules of fatty food reach the small intestine, they are grasped by the epithelial cells covering the villi, and are carried eventually within the body of the villus to this lacteal.

The lacteals pierce the muscularis mucosæ, and in the submucosa are in connection with a *plexus of lymphatic tubes and spaces*. They eventually unite with *efferent lymph-tubes*, J, which open into the lymphatics of the mesentery.

Connected with the plexus of lymphatics in the submucosa are minute *nodules of lymphoid* or *adenoid* tissue, which have unfortunately been called lymphatic *glands*. They are in no sense glands.

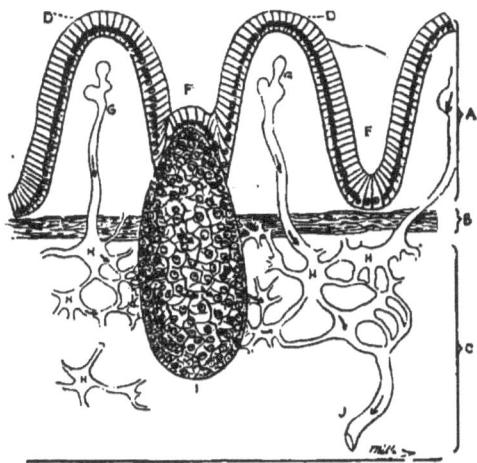

Fig. 105. Diagram showing Portions of Intestinal Mucous Membrane concerned in Absorption.
  A. Mucosa.
  B. Muscularis mucosæ.
  C. Submucosa.
  D, D. Villi.
  F, F. Crypts of Lieberkühn.
  G, G. Lacteals.
  H, H. Chinks and intercommunicating channels of the lymph-plexus of the submucosa.
  I. Bottom of a mass of adenoid or lymphoid tissue — a so-called solitary gland. Peyer's patches are formed of aggregations of these nodules.
  J. Efferent lacteal or lymph duct.

Slit up a portion of intestine along the attached border, and carefully examine the inner surface: it will present a velvety appearance, due to the minute villi. You will also find little nodules, perhaps one or two millimeters in diameter, scattered here and there in the mucous coat. These are the lymphatic nodules alluded to above—the so-called *solitary glands*. One of the nodules is indicated in the diagram at I, with its point projecting between the villi at F.

Continuing your examination of the gut, you will discover, especially in the ileum, roughened patches perhaps five to ten centimeters long by one to two centimeters broad. These are collections of the lymphatic nodules described in the last paragraph, and are termed *agminate glands*, or *patches of Peyer*. They have no secretive power, being simply in connection with, and a part of, the chain of lymphatics in the walls of the intestine. They consist of lymphoid or adenoid tissue, which will be described with the lymphatics.

To recapitulate, the small intestine presents the following:

1. The *villi*, each containing a plexus of blood-capillaries and the lymphatic or absorbent vessel.
2. *Crypts or follicles of Lieberkühn.*
3. *Brunner's glands.*
4. *Solitary lymphatic nodules*, the so-called solitary glands.
5. *Agminate lymphatic nodes*, agminate glands or patches of Peyer, consisting of aggregations of lymphatic nodules similar to the solitary lymph-follicles.

The muscular part of the intestine is arranged not unlike that portion of the stomach: *i. e.*, with an inner circular and an outer longitudinal layer. Between the two is located *Auerbach's plexus* of non-medullated nerves. A similar plexus, *Meissner's*, is found in the submucosa. The ganglia may rarely be seen in ordinary sections.

A small quantity of areolar tissue connects the external longitudinal muscular layer with the peritoneal investment.

PRACTICAL DEMONSTRATION

The intestines of the dog or rabbit are more commonly used for practical work, for reasons already alluded to. The tissue should be cut into small pieces, and hardened quickly in alcohol. When human intestine can be obtained fresh, a piece, say three inches long, should be emptied of its contents, filled with alcohol by tying the ends, and the whole hardened in strong spirit. Under no circumstances should the gut be washed, and great care must be taken to avoid injuring the delicate cells covering the villi. Vertical sections with the microtome are the most valuable. Stain with hæmatoxylin and eosin, and mount permanently in balsam.

VERTICAL SECTION OF THE ILEUM, INCLUDING PORTION OF A PATCH OF PEYER. HUMAN (*Vide* Fig. 106)

OBSERVE:

(L.)

1. The villi. (*a*) That they are of varying lengths, slender, wavy and delicate. (*b*) The covering of columnar cells. (The

FIG. 106. INTESTINAL MUCOUS MEMBRANE THROUGH A PEYER'S PATCH. VERTICAL SECTION. STAINED WITH HÆMATOXYLIN AND EOSIN (×250).

A, A. A. Villi.
B. Transverse sections of crypts of Lieberkühn.
C, C. Crypts in vertical section.
D, D, D. Nodules of lymphoid tissue—constituting a patch of Peyer.
E. Muscularis mucosæ.
F. Submucosa.

free extremities of many of the villi in the drawing are seen broken, and the epithelium is wanting in places. It is almost impossible to secure perfect villi from human intestine, on account of the length of time usually intervening between death and the removal of the tissue.) (*c*) **Oblique sections.**

2. The **crypts of Lieberkühn.**

3. The lymphatic nodules (so-called solitary glands), constituting the elements of a patch of Peyer. (*a*) **Their projection upon the mucous surface of the gut between the villi.** (*b*) **The covering with epithelium** on their free borders. (They are

located, properly speaking, in the submucosa and between the villi. In the drawing, their bases do not all appear in the submucosa, inasmuch as the nodules are cut in different planes.)

4. **Muscularis mucosae.** The **elongated nuclei** of the involuntary muscular elements.

5. The **submucosa.** (*a*) The **blood-vessels.** (*b*) **Lymph-spaces.** (Lymphatic channels are very irregular in form and size, and are often mistaken, in sections, for ruptures in the connective tissue. The stained nuclei of the endothelial cells, with which all lymph-channels are lined, will enable one to differentiate.)

(H.)

6. **The villi.** (*a*) **The covering columnar cells.** (*b*) **Goblet cells** scattered between the last. (These goblet or mucous cells are well shown in the intestine of the dog or rabbit.) (*c*) **The lacteals.** (These are not plainly demonstrable, under ordinary circumstances, in human tissue. Sections from the gut of a dog, killed during the active digestion of materials rich in hydrocarbons, will show them filled with minute fat-globules.) (*d*) The **basis tissue,** a fibrous reticulum containing many lymphoid cells. (*e*) **Portions of the capillary plexuses.**

7. **Blood-vessels of the mucosa** below the villi.

8. The **lymphoid** or **adenoid tissue** of the lymph-nodules.

Mount also sections of (1) DUODENUM, to show Brunner's glands, which occur there only. (2) LARGE INTESTINE (human), which has no villi nor valvulæ conniventes. The crypts of Lieberkühn and solitary follicles are abundant. (3) VERMIFORM APPENDIX (human), observe the abundant lymphoid tissue.

THE LIVER

This great gland is covered with a fibrous membrane—the *capsule of Glisson*. The capsule is covered with a single layer of irregularly shaped, flat endothelial cells.

Prolongations from the fibrous, visceral portion of Glisson's capsule penetrate the organ from every side, and divide the entire structure into compartments—the *lobules*.

The hepatic lobules are irregularly polygonal in transverse section, and somewhat ovoid vertically. They are about two millimeters in diameter.

Let us first examine the general plan of the vascular arrangement, and later, the minute structure of the lobular parenchyma.

The hepatic blood-supply comes from two sources: 1. The venous drainage collected in the portal vein. 2. The arterial supply, provided from the aorta by the hepatic artery. The portal venous blood is filtered through the liver instead of passing directly to the ordinary destination of venous blood (the vena cava), in order to contribute certain factors to the processes of digestion and metabolism, while the smaller arterial supply is distinctly nutritive. The *hepatic duct* is the common excretory conduit of the bile after its formation by the parenchyma of the liver.

The scheme of the organ will be understood by reference to Fig. 107, which is purely diagrammatic.

The portal vein enters the liver at the transverse fissure. It divides and subdivides; and, reaching every part of the various lobes, the terminal twigs are seen in the connective tissue of the walls of the lobules.

Branches from these termini of the portal or *interlobular veins* penetrate the lobular areas, and immediately break up into capillaries, which form an intricate plexus throughout the lobule. The blood from these capillaries is finally collected into a *central or intralobular vein*, by means of which it is immediately drained from the lobule.

The central veins from a varying number of the lobules unite outside of the latter, forming the beginning of the hepatic or so-called *sublobular veins;* and sublobular veins from various lobular areas unite, forming several (six or seven) large *hepatic veins*, which, passing in the connective tissue framework, finally drain the blood.

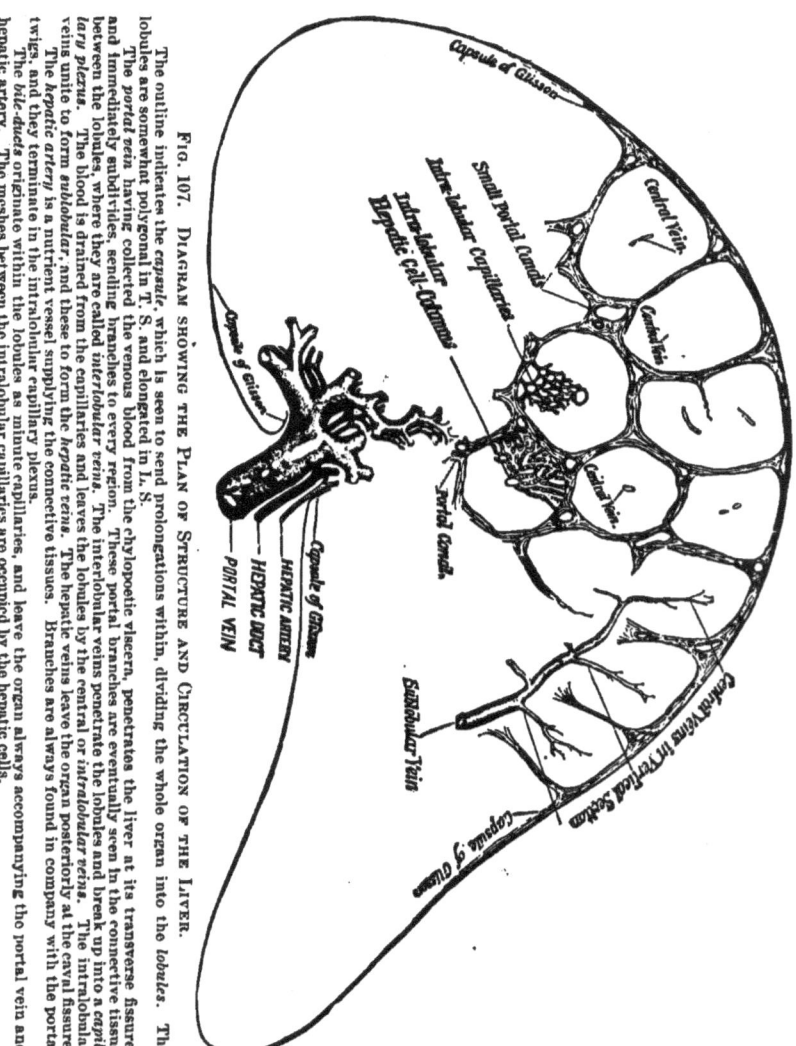

FIG. 107. DIAGRAM SHOWING THE PLAN OF STRUCTURE AND CIRCULATION OF THE LIVER. The outline indicates the *capsule*, which is seen to send prolongations within, dividing the whole organ into the *lobules*. The lobules are somewhat polygonal in T. S. and elongated in L. S.

The *portal vein*, having collected the venous blood from the chylopoetic viscera, penetrates the liver at its transverse fissure, and immediately subdivides, sending branches to every region. These portal branches are eventually seen in the connective tissue between the lobules, where they are called *interlobular veins*. The interlobular veins penetrate the lobules and break up into a *capillary plexus*. The blood is drained from the capillaries and leaves the lobules by the central or *intralobular veins*. The intralobular veins unite to form *sublobular*, and these to form the *hepatic veins*. The hepatic veins leave the organ posteriorly at the caval fissure.

The *hepatic artery* is a nutrient vessel supplying the connective tissues. Branches are always found in company with the portal twigs, and they terminate in the intralobular capillary plexus.

The *bile-ducts* originate within the lobules as minute capillaries, and leave the organ always accompanying the portal vein and hepatic artery. The meshes between the intralobular capillaries are occupied by the hepatic cells.

from the organ and pour it into the ascending cava as it lies posteriorly in its fissure.

The *hepatic artery* also penetrates the transverse fissure. It accompanies the portal vein in its ramifications, giving off *nutrient twigs to the connective tissue framework* and to the *walls of the vessels*. The terminal branches, very minute, pour any remaining blood into the venous plexus at the margin of the lobules, thus providing arterial blood for the lobular parenchyma.

The *hepatic duct* is also seen emerging from the transverse fissure. (For the sake of clearness, we will trace it from without inward.) It follows the course of the portal vein with the hepatic artery. Wherever in a section of the organ the portal is divided, the artery and duct will also appear. Bound together with connective tissue, the trio reach the walls of the lobules. The ducts now penetrate the lobule and break up into an exceedingly minute plexus—the *bile-capillaries*. This plexus properly *begins in the lobules* and drains the bile as formed, passing it into the ducts in a direction opposite to the portal blood-current.

THE PORTAL CANALS

If it were possible to grasp the vessels as they are found emerging at the transverse fissure, the portal vein, hepatic artery, and hepatic duct, and to forcibly tear them, with their supporting connective tissue, out of the liver, a series of channels or canals would thereby be formed. *A portal canal, then, is a space in the liver occupied by branches of the portal vein, the hepatic artery, and the hepatic duct, and the contiguous connective tissue.* Frequently more than one specimen of each vessel is to be seen in a canal. There may be two or three veins, and as many arteries and ducts, associated in a single portal canal. Lymphatic chinks are also abundant in this connective tissue.

From what has been said, it will be understood that *a vessel found by itself* in this organ *must be either an intralobular or a hepatic vein;* and these are easily distinguished, as the former are within, while the *latter are without the lobules* and in the connective tissue framework. On the other hand, a *group of vessels will indicate a portal canal*, with its large and thin-walled vein, the small thick-walled artery, and, intermediate in size, the duct.

THE LOBULAR PARENCHYMA

The lobules consist of two capillary plexuses, one containing blood and the other bile. In the meshes of this network, the hepatic cells are located.

The blood-capillaries, although extremely tortuous, have a general direction of convergence toward the central veins. This is best seen when the lobules have been divided in a vertical direction.

The bile-capillaries are among the smallest canals found in vascular tissues, having a diameter of only 1 to 2 µ. They pursue a direction in the human liver, as a rule, at *right angles to the course of the blood-capillaries*, and are not demonstrable, except with considerable amplification, say × 400, and then only in the thinnest portion of the sections. They are, properly speaking, merely minute channels in the parenchyma, and have, it is believed, no wall.

The hepatic cells are polyhedral, about twice the size of a white blood-corpuscle, say from 20 to 25 µ in diameter, usually with a single nucleus and with granular protoplasm, frequently contain-

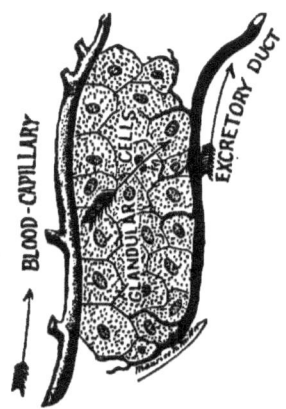

FIG. 108. DIAGRAM ILLUSTRATING THE INTRALOBULAR HISTOLOGY OF THE LIVER.

The hepatic cells are connected in columns between the blood-capillaries. The cells are endowed with the power of selecting especially such materials from the blood as are necessary for the manufacture of bile. Having accomplished this, the secreted fluid is given up to the bile-capillaries, and by them poured into the ducts, and led out of the liver for subsequent use. The direction of the pressure is indicated by the arrows. This is the histology of gland-structures generally.

ing minute fat-droplets and granules of yellow pigment. The existence of a definite limiting membrane has been questioned, as far as the cell of the human liver is concerned, although such a structure can be shown in many of the lower animals.

The physiological plan of the intralobular structure is expressed in the diagram, Fig. 108. The blood is brought into relation with the lobular parenchyma—the hepatic cells—by the capillary plexus, and the elements necessary to constitute the bile are selected and carried on, to be drained away by the bile-capillaries and-ducts.

PRACTICAL DEMONSTRATION

It is best to begin with the liver from a pig. The amount of connective tissue in the normal human liver is very small, and is mainly confined to the support of the interlobular vessels; the boundaries of the lobules are, therefore, poorly defined, and without the previous observation of some well outlined specimen the student frequently gets but an imperfect notion of the plan of the human organ.

Pieces of liver, say a centimeter square by half a centimeter thick, are hardened by twenty-four hours' immersion in strong alcohol. Larger pieces may be prepared with Müller's fluid. Sections should be cut with a microtome, care being taken to include some of the medium-sized portal canals. The portal vein, with its accompanying vessels, may be easily distinguished from the solitary and less frequent branches of the hepatic veins. The elements of these canals, and especially the larger ones, are best kept intact by infiltration of the tissue with celloidin; but very fine sections may, with care, be made from the alcohol-hardened tissue. Even free-hand cuts, after some degree of skill has been obtained by practice, will answer very satisfactorily. Stain with hæmatoxylin and eosin.

SECTION OF LIVER OF PIG. CUT VERTICALLY TO AND INCLUDING THE CAPSULE OF GLISSON

OBSERVE: (Fig. 109)

(L.)

1. The **capsule of Glisson**, C. (Note the prolongations sent into the organ, which divide the entire structure into irregularly polygonal areas—if divided transversely—and elongated, vertically-sectioned areas—the hepatic lobules.)

2. The **central (intralobular) veins**, C. V. (Note that the figure formed by the division of the vein varies according to the direction of the cut, a circle, oval, or elongated slit, as the lobules have been sectioned transversely, obliquely, or vertically.)

3. The **hepatic veins**, H. V. (Those shown in the section are undoubtedly sublobular. It must be remembered that *sub* applied to these vessels is misleading, as the lobules are situated on every side, as well as above the sublobular veins.)

4. The **portal canals**, P. C. (Even the smaller ones, I, I, are readily differentiated from areas containing hepatic veins, inasmuch as a *group* of vessels can be distinguished—the hepatic veins running alone.)

168 STUDENTS HISTOLOGY

5. The **portal veins**, V. (Observe that they usually are the largest vessels in the canals. Note their thin walls. They not infrequently contain blood-clots, with deeply stained, scattering,

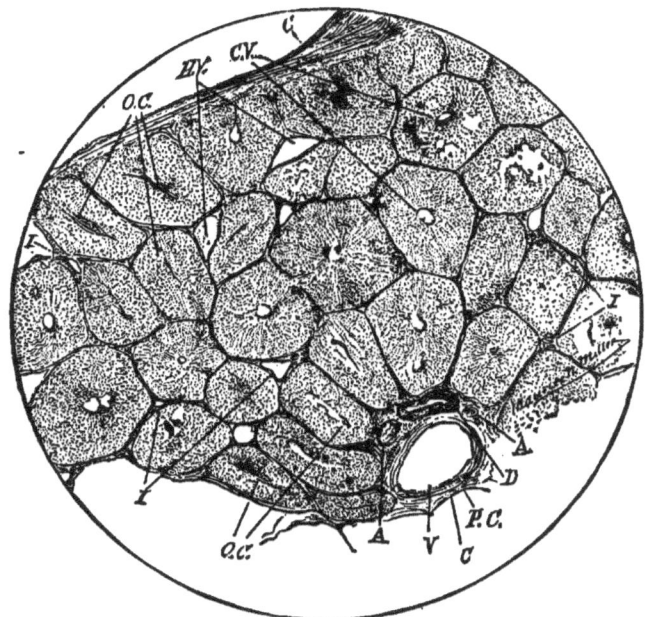

FIG. 109. LIVER OF THE PIG SECTIONED AT A RIGHT ANGLE TO GLISSON'S CAPSULE. STAINED WITH HÆMATOXYLIN AND EOSIN (× 60).

C. Capsule of Glisson.
C. V. Central or intralobular veins.
O. C. Oblique section of central veins.
I. I, I. Interlobular veins. (In small portal canals.)
P. C. A large portal canal.
A. A. Hepatic arteries.
D. Hepatic duct.
V. A portal vein.
H. V. Hepatic veins—probably sublobular.

white corpuscles, appearing with this amplification as dots or granules.)

6. **Hepatic arteries**, A. (The larger examples may be determined by their thick muscular media and the wavy pink line—the fenestrated membrane. Several may be seen in a single canal.)

7. **Hepatic ducts**, D. (These are lined with cylindrical epithelial cells, hexagonal in transverse section, and the bold, deeply

stained nuclei give the ducts marked prominence even with the low-power. Indeed, the smaller portal canals are frequently differentiated by this element alone—this being especially true when the structures have been disturbed, and perhaps torn, in the process of mounting.)

8. The **lobular parenchyma**. (The arrangement of the hepatic cells, forming branching columns, is merely indicated—with the low-power—by their deeply stained nuclei presenting granular areas within the lobular boundaries. Still, by careful attention, the elements will be seen to radiate more or less distinctly from focal points—the central or intralobular veins.

(**H.**)

9. The **portal veins**.

10. The **lymph-spaces** in the connective tissue of the portal canals. (Note, in those which are better defined, the nuclei of the endothelium. Do not confound these lymphatics with small veins, as the latter present a tolerably defined wall, while the lymphatic chinks appear like rifts in the connective tissue; it would be difficult to make this distinction without the endothelial cells.)

11. **Hepatic arteries.** (On account of its solidity, the liver will enable the student to secure sections of blood-vessels presenting the typical structure more nearly than the specimens obtained from the organs heretofore examined.) Note (a) the **elongated nuclei of the muscular elements of the media**; (b) the **fusing of the adventitia** with the connective tissue surrounding the artery; (c) the sharply defined **outer boundary of the intima**—the **fenestrated membrane**, which, from the action of the hardening agent, has contracted the elastic fibers and detached (d) the **endothelial cells**. (Inasmuch as the lining cells of small arteries are very frequently partly detached in alcohol-hardened tissue, they may simulate columnar cells. A like appearance is often presented when an artery has been sectioned obliquely, by the projecting muscle-cells of the media.)

12. **Hepatic ducts.** Note : (a) The **lining cylindrical cells.** (b) The **nuclei** of these cells (as a rule, perfectly spherical ; and, in transections arranged in a circle, affording an appearance perfectly characteristic). (c) **Mucous glands** in the wall of the larger ducts, lined with large, nucleated, columnar cells, precisely like those lining the duct-lumen ; and, hence, liable to be mistaken for small ducts. (The tube carrying the mucus secreted in

these pocket-like glands does not pass directly into the lumen of the duct, but runs along obliquely, much like glands in the bronchi. Not infrequently the glands possess no proper efferent tube, but are mere depressions or diverticula in the thick wall of the bile-duct.)

13. The **lobular parenchyma**. (Single cells, partly detached, may be found about the edges of the section.) Note: (*a*) The somewhat **polygonal figure**; (*b*) the **nucleus**; (*c*) **nucleoli**; (*d*) **fibrillated**, mesh-like **cell-body**; and (*e*) an apparent **cell-wall**. (The arrangement of the lobular parenchyma will be noted in connection with the human liver.)

HUMAN LIVER

PRACTICAL DEMONSTRATION

The sections from which the illustrations have been drawn were made from material hardened in Müller's fluid. The tissue was then cut, the sections washed by six hours' maceration in water, after which they were treated successively with weak and stronger alcohol, stained with hæmatoxylin and eosin, and mounted in balsam. This treatment aids greatly in the demonstration of the blood-capillaries, as the contained blood-corpuscles, in consequence of some change affected by the chromium salt, take the eosin deeply. The nuclei of cells are also rendered markedly prominent.

Pieces of tissue, one centimeter square by half a centimeter thick, may be hardened in alcohol. This method will give very excellent results, providing the sections be cut as soon as the hardening process has become complete. Stain as above.

For the demonstration of the isolated hepatic cells, scrape the cut surface of a piece of hardened liver with a scalpel, and throw the scrapings into a watch-glass of hæmatoxylin. After a few moments drain off the stain, and brush the stained tissue elements into a test-tube nearly filled with water. Change the water two or three times; and when clear, add a few drops of eosin solution. Allow the eosin to stain for a moment only; decant, drain, and fill the tube with alcohol. After ten minutes the spirit may be drained off and the tube partly filled with oil of cloves. A drop of the sediment may then be placed upon the slide, the bulk of the oil removed with paper, and the mounting completed by adding a drop of balsam and the cover-glass. This tissue may be kept in the oil from year to year for class-room purposes. If the oil be pure and the washing thorough, the staining will remain unaffected for two or three years.

SECTION OF HUMAN LIVER, CUT AT A RIGHT ANGLE TO THE SURFACE, AND STAINED WITH HÆMATOXYLIN AND EOSIN
(Fig. 110)

OBSERVE:

(**L.**)

1. The **imperfectly outlined lobules** (in consequence of the absence of interlobular connective tissue.)

2. The **fusing of the lobules.** (At points like B, B, it is impossible to say just where one lobule ceases and the contiguous one begins.)

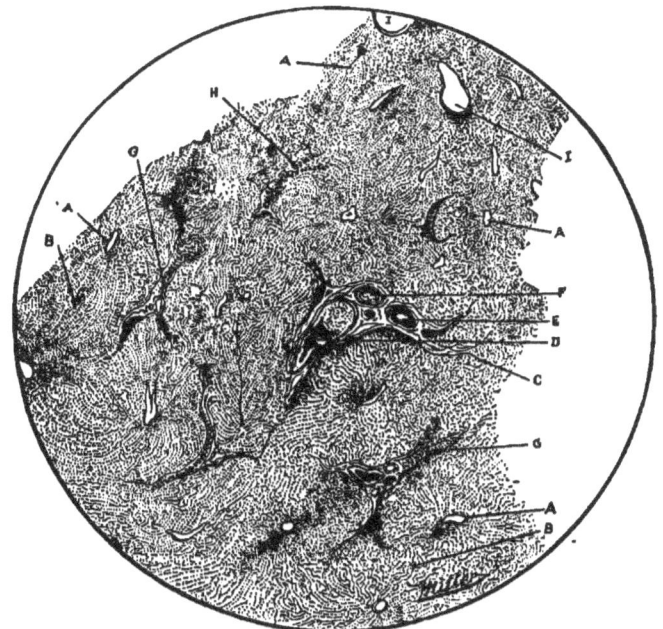

FIG. 110. SECTION OF HUMAN LIVER. STAINED WITH HÆMATOXYLIN AND EOSIN (×60).

A, A, A. Central veins sectioned generally at a right angle to the lobule.

B, B. Points where adjoining lobules coalesce. Illustrating the difficulty of outlining the lobules in normal human liver.

C. Connective tissue of a portal canal.

D. Large interlobular vein.

E. Hepatic duct belonging to C.

F. Hepatic artery of C.

G, G. Smaller portal canals.

H. Small hepatic ducts—always recognizable by the deeply hæmatoxylin-stained nuclei of their lining cells.

I, I. Hepatic—sublobular—veins.

3. The **central** (or intralobular) **veins**, A, A (frequently appearing as mere slits on account of the direction of the cut).

The **portal canals**, G, G. (These are readily detected on account of the deeply stained nuclei of the cells lining the hepatic ducts.) (**H.**)

5. **Portal canals** (too small for demonstration of the several elements, but always distinguishable by the columnar epithelial cells.)

6. The **larger portal canals**, C. Note: (*a*) The large thin-walled **vein**, D; (*b*) The **duct**, E; (*c*) The **artery**, F.

7. The tortuous course of the **hepatic cell-columns** as compared with the same in the section previously studied.

8. The **hepatic veins**. (Observe their infrequency compared with the sections of the portal veins. Note the small amount of connective tissue around them—greater, however, than that about the central veins.)

ELEMENTS OF A PORTAL CANAL—FROM PREVIOUS SECTION

OBSERVE: (Fig. 111)

(**H.**)

1. The **portal vein**, V. (Note the nuclei of the few **endothelial cells** remaining, and the **corpuscular elements of the blood** in the lumen of the vein. Observe that the white corpuscles are scanty, and deeply stained; also that many of the colored corpuscles are granular, and show loss of **pigment** from action of the alcohol.)

2. The **hepatic artery**, A. (In the human liver, the portal canals frequently carry a number of arteries and ducts, instead of one of each, as shown in the one selected for the illustration. The arteries can nearly always be differentiated by the clear, wavy line of the fenestrated membrane. Should the section have been in a longitudinal direction with reference to the vessel, look for the elongated nuclei of ·the unstriated muscle-cells of the media, some running with the artery—the longitudinal—and others at right angles to its course—the circular fibers.)

3. The **hepatic duct**, D. (Observe the **thickness of the wall**, depending, of course, upon the diameter of the duct itself, and the presence of **connective tissue** supporting scattering **unstriated muscle-cells**. Note the beautiful, clear, columnar cell-lining. That these epithelial cells are **polygonal in transverse section** is demonstrable at D, L, where the duct has been cut in a longitudinal way, and the cells are seen from above.

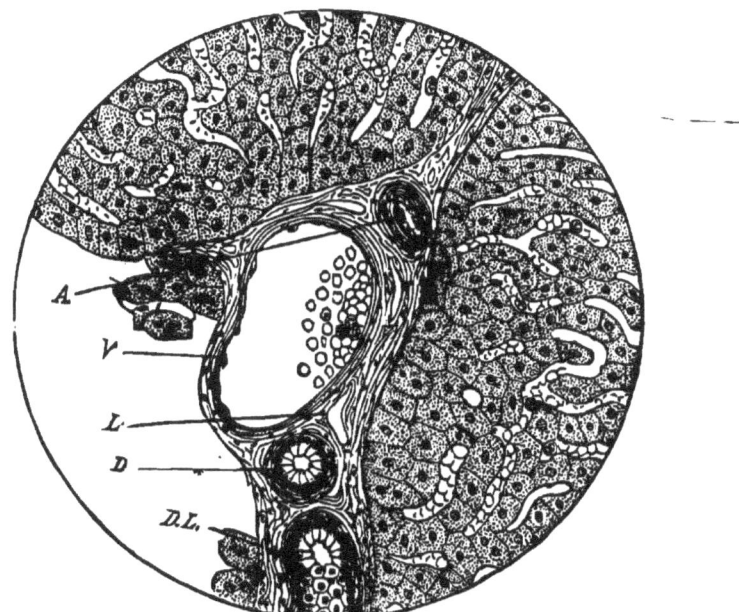

FIG. 111. SECTION OF HUMAN LIVER SHOWING THE ELEMENTS OF A PORTAL CANAL. STAINED WITH HÆMATOXYLIN AND EOSIN (×400).

A. Hepatic artery.
V. Portal vein—interlobular.
D. Hepatic duct in T. S.
D, L. Hepatic duct in L. S.
L. Lymph-space.

The lobular parenchyma of contiguous lobules will be seen on the right, and above the canal.

4. The **connective tissue elements** of the canal, reaching out in various directions between the adjacent lobules.

5. **Lymph-spaces** or -chinks, L. (Note the stained nuclei of the endothelial cells.)

6. **Nerve-trunks.** (In the larger canals bundles of non-medullated or more rarely medullated nerve-fibers may be frequently seen. They are not shown in the accompanying illustration.)

THE LOBULAR PARENCHYMA—STAINED CELLS FROM HUMAN AND PIG'S LIVER (Fig. 112)

OBSERVE:

(H.)

1. **Isolated hepatic cells** A, A. Note the **large, variably-sized nuclei**, their nucleoli, and the **granular protoplasm** of the cell-body.

2. **Groups of cells** forming portions of the hepatic cell-columns, as at C.

3. **Cells containing fat-globules,** D.

4. **Glycogen** can be demonstrated in the hepatic cells of the rabbit some hours after a meal of carrots. Harden in alcohol; cut without imbedding;. stain with iodine solution (see page 29), which colors glycogen reddish brown.

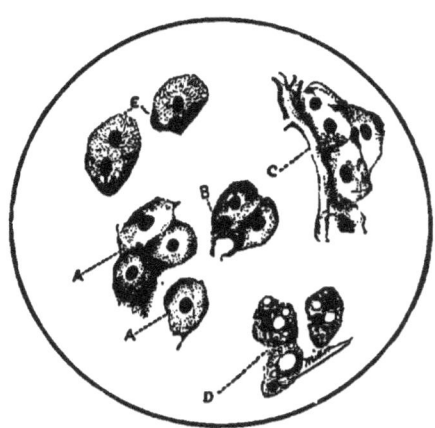

FIG. 112. ISOLATED HEPATIC CELLS. STAINED WITH HÆMATOXYLIN AND EOSIN ($\times$ 400).

A, A. Cells from human liver.
B. Cells from same, showing below a blood-capillary in T. S.
C. A blood-capillary with part of a column of cells.
D. Human liver cells in a condition of fatty infiltration.
E Isolated cells from liver of pig, showing intracellular network.

THE LOBULAR PARENCHYMA, CONTINUED—SECTION OF HUMAN LIVER (Fig. 113)

Having found with L a typical lobule in transverse section:

OBSERVE :

(H.)

1. The **central vein,** C. V. (Note the exceedingly delicate wall, and search for a trunk of the intralobular plexus in its connection with this vein.)

2. The **blood-capillaries in longitudinal section,** B. C. (Observe their **tortuousness, bifurcations,** and **anastomoses.**)

3. **Blood-capillaries in transection,** T. S. (Should the

capillaries be filled with blood, this demonstration will be greatly aided.)

4. **Hepatic cell columns**, H. C. (Note the difficulty with which these can be traced for any great distance, on account of their **irregular and twisted course** throughout the lobule. Observe that the lobules are composed largely of tortuous blood-capillaries, between which the hepatic cell-columns are placed. Note the manner in which the **cells are disposed around the blood capillaries**, as at T. S.)

5. **Bile-capillaries**, D. (These are rather difficult of demonstration in the human liver. The section should be extremely

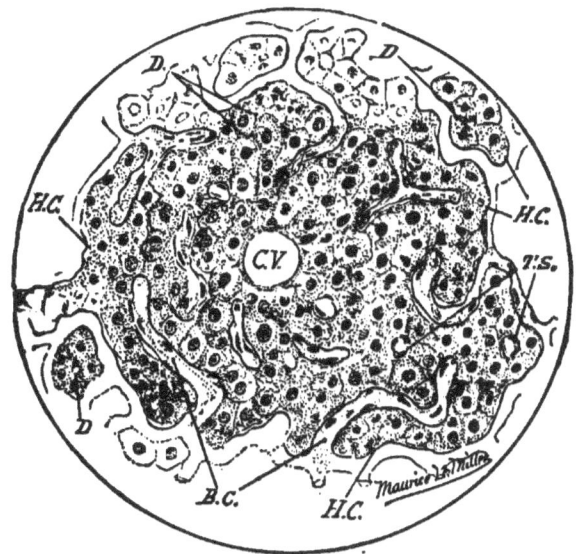

FIG. 113. A SINGLE LOBULE FROM HUMAN LIVER. TRANSVERSE SECTION. STAINED WITH HÆMATOXYLIN AND EOSIN ($\times$ 400).

C. V. Central vein of the lobule.
B. C. Blood-capillaries in L. S.
T. S. The same in transverse section.
H. C. Columns of hepatic cells.
D. Bile-capillaries.

thin, and, in order to get the best results, a higher power instrument than we ordinarily use will be required. The best point at which to see them is at the junction of three or four cells, where the bile-capillary has been divided transversely.

THE LOBULAR PARENCHYMA, CONCLUDED—ORIGIN OF THE BILE-
DUCTS—SAME SECTION AS BEFORE   (Fig. 114)

OBSERVE:

(H.)

1. The **connection between the intralobular bile-capillaries and the marginal or intralobular bile-ducts.** (The manner of connection between the above is as follows: The bile-capillaries

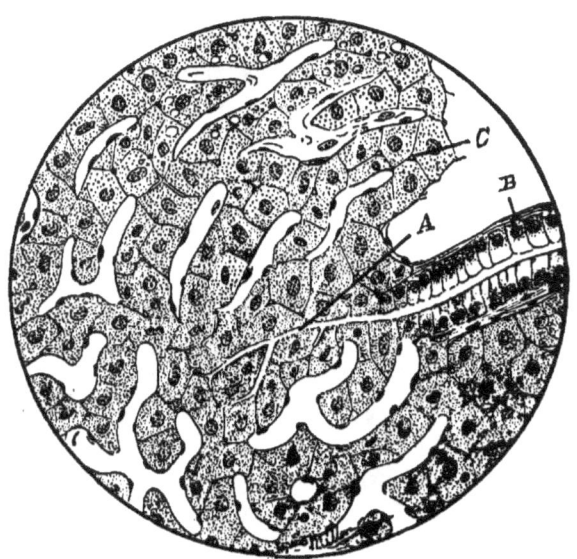

FIG. 114. PORTION OF THE PERIPHERY OF A HEPATIC LOBULE SHOWING THE ORIGIN OF A BILE-DUCT. STAINED WITH HÆMATOXYLIN AND EOSIN (×400).

A. Bile-capillaries in longitudinal section.
B. Bile-duct. The bile-capillaries are simply chinks between the hepatic cells. In the formation of a duct, the hepatic cells become altered in shape, elongated, and eventually become the lining cells of the duct. A little connective tissue, thrown around the outside, completes the structure as seen at B.
C. Bile-capillary in transverse section. The larger clear spaces are blood-capillaries.

are merely channels between the hepatic cells, and run, as a rule, at a right angle to the blood-capillaries. They are probably destitute of a wall in the human liver. As these channels approach the marginal part of the lobule, the hepatic cells surrounding the capillary are seen to change their form. *They elongate, becoming thinner, gradually losing their form as hepatic cells, and assume a columnar type. At the same time, a few fibers of con-*

nective tissue are thrown outside the modified hepatic cells, and a bile-duct results. The hepatic cells become, insensibly, the columnar cells lining the duct. This is shown in the illustration rather diagrammatically. Its demonstration requires much patient study and search. The duct is best traced backwards to the point where the bile-capillaries enter it.

The study of the blood- and bile-capillaries of the liver is much easier if tissues are used in which these vessels have been *injected* with some colored substance. Injection of the bile-capillaries is difficult; but injection of the blood-capillaries is quite readily accomplished.

The structure of the *gall-bladder* and its duct is substantially the same as that of the larger bile-ducts. The mucous membrane is thrown into numerous small folds, and is covered by columnar epithelial cells. The mucous membrane is supplemented by unstriated muscle-fibers, outside of which is a fibrous layer.

THE KIDNEY

The kidney is as singular in structure as in function. Although as originally developed it is divided into lobules, little trace of this structure remains in the adult organ.

The kidney consists, essentially, of an intricate system of blood-vessel plexuses, in intimate relation with a system of urine-tubes—the whole supported by a small amount of connective tissue.

The accompanying drawing (Fig. 115) will serve to give an idea of the gross plan or scheme of the structure—remembering that the illustration is only a *diagram*.

On making a vertico-lateral section, on the median line, the following parts are seen:

The kidney is invested with a *fibrous capsule*, which is connected with the parenchyma by very delicate prolongations of its connective tissue fibrillæ. This capsular investment is in connection above with the supra-renal bodies, and, on the inner border, with the vessels, etc., which enter and leave the organ at its hilum. The ureter, penetrating the areolar tissue which (containing much fat) surrounds the hilum, may, for clearness of description, be traced backward into the kidney. This tube expands into the pelvis, and reduplications of its wall imperfectly divide the pelvis into three compartments, or *infundibula*.

Each infundibulum is subdivided again, imperfectly, into several pockets or *calyces;* and into each calyx may be seen, peeping from the kidney-substance, a *papillary eminence* or apex of a cone —the *pyramids of Malpighi*. The pelvis is lined with a variety of transitional or imperfectly stratified epithelium, which will be described hereafter.

The blood-vessels, lymphatics, etc., pass in at the hilum, outside the ureter, pelvis, and infundibula. The artery divides into numerous branches, which are seen in the diagram passing outward, between the Malpighian pyramids. The renal vein pursues much the same course, the main trunks lying side by side.

On examining a section of the kidney, made in the direction indicated in Fig. 115, a division of an outer portion will be manifest, bounded externally by the capsule of granular texture, containing the blood-vessels, etc. This is called the *cortex*. Within the cortical portion there appear a number of pyramidal masses—whose apices we have previously seen—of finely striated texture—the medullary or *Malpighian pyramids*. The cortical substance projects itself between the pyramids, completely isolating them, and forming the *cortical columns* of Bertini.

Again observing the outer cortex, it will present narrow, light-

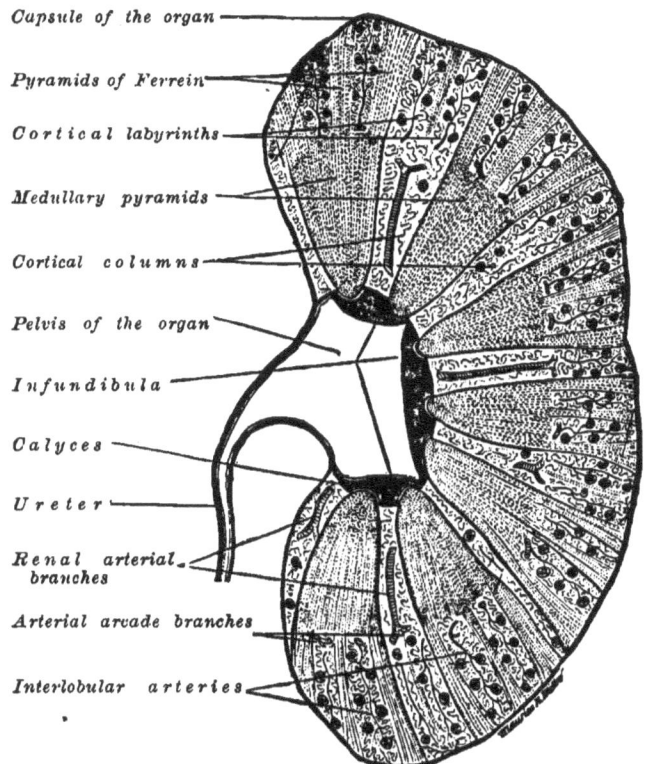

Fig. 115. Diagram Showing the Plan of Structure of the Human Kidney.

colored lines, which converge toward the pelvis; and, eventually, pass into and become a part of the Malpighian pyramids. These light areas, made up of tubules, are the *pyramids of Ferrein*, or, as sometimes called, the medullary rays.

The darker spaces between the pyramids of Ferrein are called *labyrinths*.

The gross elements, to be understood before we proceed any further, then, are:

1. The *capsule* of the kidney.
2. The *ureter*.
3. The *pelvis*, with its three infundibula. The subdivision of each infundibulum into several calyces. Each calyx the site of the apex of a Malpighian pyramid.
4. The *blood-vessels* entering and leaving the hilum. Their subdivision outside the pelvic lining, and final passage into the kidney-substance in the cortical columns.
5. Division of kidney-substance into *cortex* and *medullary* or *Malpighian pyramids*.
6. Penetration of cortical tissue inward between pyramids of Malpighi—constituting the *cortical columns*.
7. The *pyramids of Ferrein*.
8. The *labyrinths*.

In the domestic animals there are no cortical columns—the pyramids of Malpighi coalescing, as it were—thus presenting a true medulla.

We have remarked that the kidney is made up largely of urine-carrying vessels (the tubuli uriniferi) and blood-vessels. We will first study the tubular system, reserving for the present the consideration of the blood-vessel arrangement.

THE TUBULI URINIFERI

The urine-carrying tubules commence in the cortex, and, after taking a very circuitous route with frequently varying diameters, end at the apex of the pyramids of Malpighi, where they pour their contained urine into the calyces. The urine then overflows into the infundibula, and is finally drained from the pelvis by the ureter.

We shall begin with a single typical tube; and, understanding its histology, we can build up the organ by simply multiplying this element.

A uriniferous tube, or tubule, commences in the cortex in the labyrinth (between the pyramids of Ferrein), as a thin-walled sac .2 mm. in diameter. This vesicle, with its contents, is a *Malpighian body;* and its wall is called the *capsule* of the same, or the *capsule of Bowman*. The capsule consists of a thin layer of connective tissue lined by flat epithelial cells.

From one side of this, the expanded beginning of the tube, a narrow neck (25μ in diameter) is projected, which immediately widens (50 μ) into a tube—the *proximal convoluted tubule*. This tube (or this portion of the tube) pursues a very tortuous course, always keeping between Ferrein's pyramids, and finally approaches the base of a Malpighian pyramid. Here it assumes an irregular spiral form—the *spiral tubule* (42 μ).

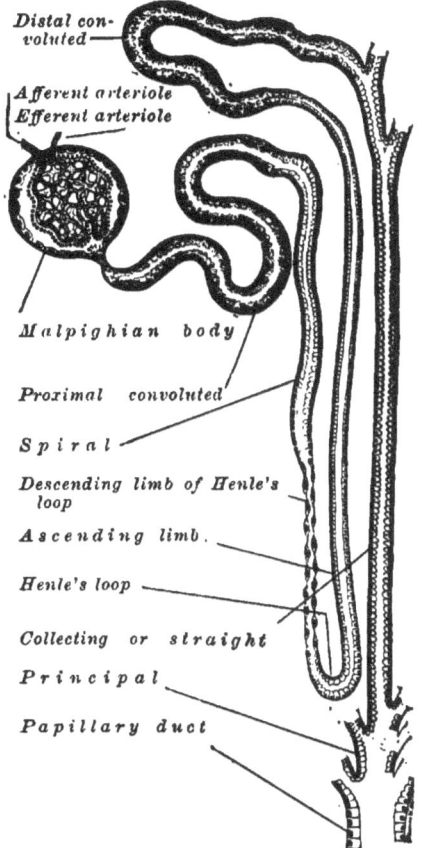

FIG. 116. DIAGRAM SHOWING THE DIVISIONS OF A KIDNEY TUBULE.

The tube suddenly narrows (10 μ), becomes straight, and passes into a pyramid of Malpighi. It reaches sometimes just into the pyramid, more frequently, however, passing deeper than this—often descending two-thirds of the distance to the apex; and is called the *descending limb of Henle's loop*. The descending limb of Henle's loop suddenly turns upon itself, forming the *loop of*

Henle; and, widening (25μ), returns upon its course as the *ascending limb of Henle's loop.* It again enters the cortex, keeping in a pyramid of Ferrein, and passes outward until it approaches the outer limit of the cortex, near the capsule of the kidney. Here the ascending limb widens (50μ), forming the *distal convoluted tubule,* which pursues a tortuous course in the outer cortex. Many histologists also recognize an *irregular tubule,* which pursues a zigzag course for a short distance between the ascending limb of Henle's loop and the distal convoluted tubule. The distal convoluted tubule then reënters a pyramid of Ferrein, narrows (40μ), and passes a second time into a Malpighian pyramid, under the title of *straight* or *collecting tube,* or *tube of Bellini.* The last, after reaching very nearly to the apex of the pyramid, unites with others of a like character to form a *principal tube* (85μ). Several principal tubes unite to form a *papillary duct* (250μ). From one hundred to two hundred of the last open upon the surface of the apical portion of a Malpighian pyramid.

It must be borne in mind that, in describing the tubular system, although such terms as "convoluted tube," "looped tube," etc., are employed, these are not separate tubes, but only *names applied to different portions of one long tube.* A single tubule, as we have seen, commences at Bowman's capsule, becomes narrowed like the neck of a flask; courses as the proximal convoluted and spiral; descends into, turns, and emerges from a Malpighian pyramid, as Henle's looped portion; reaches the extreme cortex, and swells as the distal convoluted; and here ends as a single or isolated tubule and enters a straight tube. The straight tubes receive several distal convoluted termini, at the cortical periphery, and pass in small bundles (forming the pyramids of Ferrein) directly onward toward the apex of a Malpighian pyramid, uniting with one another at very acute angles, and the trunks formed by this union uniting until the tube terminates as a papillary duct.

The tubes are lined with epithelial cells, and these cell-elements constitute the *parenchyma of the kidney.* The lining cells are, *as a rule, columnar or cuboidal.* Two exceptions are presented, one of which appears in the flattened cells lining *Bowman's capsule,* and the other in the flattened cells of the *descending limb of Henle's loop.* The parenchyma will receive attention in our practical work.

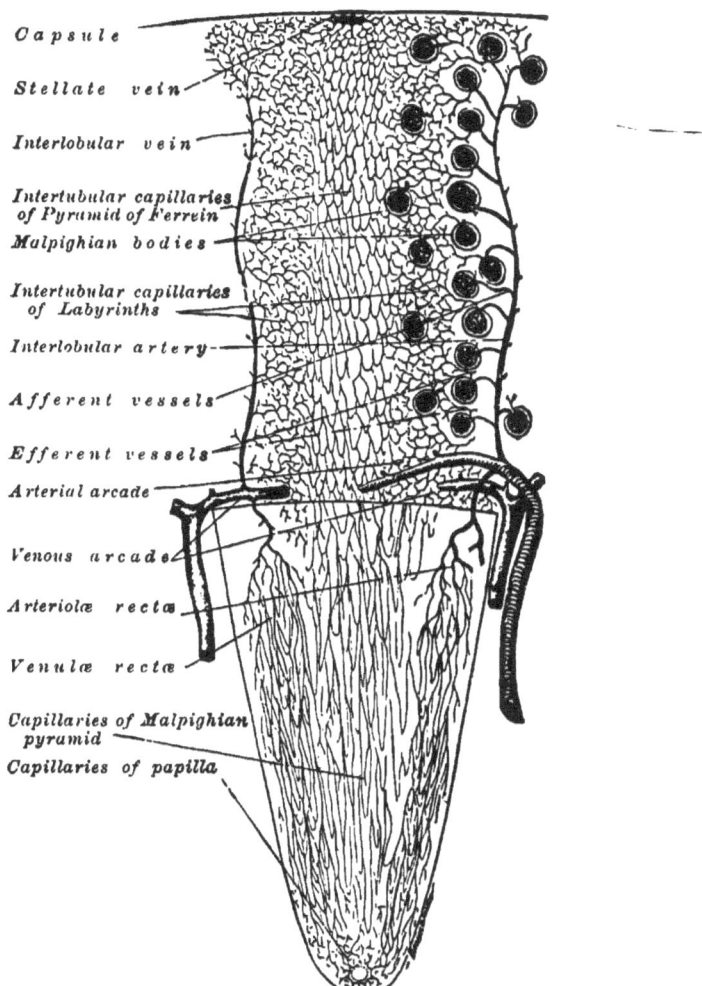

Fig. 117. Diagram Showing the Arrangement of Blood-Vessels in the Kidney—after Ludwig.

BLOOD-VESSELS

The vascular arrangement is complex. The most prominent and essential feature is afforded in the existence of two distinct capillary plexuses.

The renal artery, as already described, sends branches into the substance of the kidney. These pass between the Malpighian pyramids, and *in the cortical columns.* These arterial trunks arch over the bases of the pyramids of Malpighi, forming the *arterial arcade.* From these arches small, straight branches are sent outward toward the capsule of the kidney, occupying a position midway between the pyramids of Ferrein, in the labyrinths. The last are the *interlobular arteries.* During their course, they send off side *arterioles*, which penetrate the capsule of the Malpighian bodies. Each afferent arteriole breaks up into a fine plexus—the *tuft* or *glomerulus.* The glomerulus does not entirely fill the capsule, so that a space remains between the spherical mass of capillaries and the flattened cells lining the body. The glomeruli are enveloped with a single layer of flattened epithelial cells.

The blood escapes from the glomerulus by a minute *efferent arteriole* which emerges from the capsule close to the afferent vessel. The latter is the more noticeable, as it is usually much the larger. The efferent arteriole immediately breaks up into a capillary plexus, which courses between the uriniferous tubules of the labyrinths and of the pyramids of Ferrein. This plexus also descends between the elements of the pyramids of Malpighi. From the arteries forming the arcade another set of branches— the *arteriolæ rectæ*—is given off, which, descending into the Malpighian pyramids, provides another and direct arterial supply to the tubular elements by elongated capillary loops.

The course of the venous trunks is not unlike that pursued by the arteries. Interlobular veins, lying in the cortical labyrinths parallel with and close to the arteries, pass into a venous arcade. In the medulla the venous blood is collected from the capillaries and carried to the bases of the Malpighian pyramids in small veins—*venulæ rectæ.* The blood from the cortical intertubular capillaries is collected in the interlobular veins.

A peculiar vascular arrangement exists just beneath the capsule of the kidney, consisting of scattered venous plexuses, the *venæ stellatæ.* They contain blood collected from contiguous intertubular capillaries, and are in connection with the summits of the interlobular veins.

From what has been said, it will be seen that the cortical and medullary blood-supplies are, to a certain extent, independent of each other. The arteriolæ rectæ provide a vascular supply to the elements of the Malpighian pyramids even after many of the glomeruli have become obliterated by disease.

Nerve and lymphatic elements are not very prominent features in sections of the kidney. Small non-medullated nerve-trunks may be demonstrated in transverse sections of the cortex, especially near the bases of the medullary pyramids, where they will be seen in company with the blood-vessels of the arcades. Lymph-channels are also to be seen in the vicinity of the vessels of the hilum, and in the connective tissue of the capsule.

The histology of the kidney will be comprehended better by a reference to its function. The separation from the blood of a quantity of water, together with certain excrementitious matters, is effected, partly in the Malpighian bodies, and partly in the tubules. The vascular tuft—the glomerulus—is covered with a close-fitting membrane composed of flat cells. The blood in this plexus parts with a certain amount of its water, which passes through the walls of the capillaries and through the cells covering them. Whether this be due to osmosis or to some selective power of the cells we have no concern—suffice it to say that certain salts afterward appearing in the urine do not leave the blood at this point. The efferent glomerular arteriole, it will be remembered, breaks into a capillary plexus, which brings the blood close to the walls of the convoluted tubules. It is believed that the cells lining these tubules select from the blood circulating in the contiguous capillaries such effete materials as escaped elimination from the glomeruli.

Moreover, it seems that some of the water which escaped in the first instance and entered the proximal convoluted tubules, is here returned to the blood by the intervention of the same tubular lining cells which excrete the salts. Without referring to any further work done by the kidney, it is important to understand this part of the structural scheme: That the first part of the uriniferous tubule is the prominent excreting part. That the latter portion of the tubule—the portion in the Malpighian pyramids, the straight tubule—is for the collection and drainage of the urine already excreted. *And that between the excreting first part and the draining second part, there exists a narrow looped tubule— the loop of Henle.* The effect of this narrowing and tortuosity of the tubule will be to present a resistance to the outflow of urine from the proximal portion of the tubule. The diluted urine, excreted in the Malpighian bodies, is held back for a while in the proximal convoluted tubule, and time given for the completion and perfection of the excretory processes by the tubular parenchyma.

PRACTICAL DEMONSTRATION

The human kidney is rarely found in a perfectly normal condition. The demonstration can be made from the kidney of the pig, except as regards certain features. The medullary pyramids do not exist in the domestic animals, and the parenchyma presents very essential differences from the cells of the human kidney. Still, very much can be learned from the organ of the pig, dog, or rabbit. The tissue should be divided so as to permit sections to be made parallel with the medulla, and to include both it and the cortex. The hardening is best by Müller's fluid. Small pieces hardened quickly in strong alcohol, however, stain very finely with hæmatoxylin and eosin. Sections of kidneys the blood-vessels of which have been injected should also be studied.

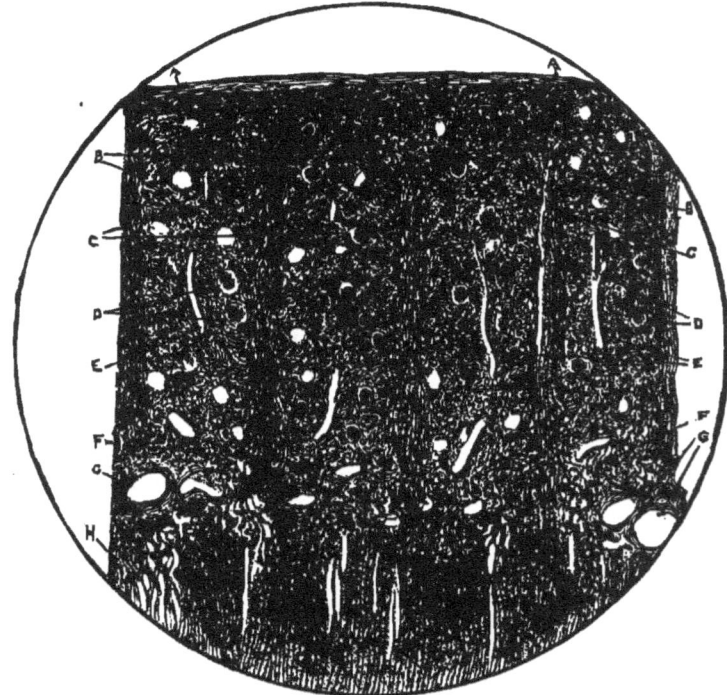

FIG. 118. SECTION OF HUMAN KIDNEY, CUT PARALLEL TO THE PYRAMIDS OF FERREIN. SHOWING THE CORTEX AND PART OF A MALPIGHIAN PYRAMID ($\times$ 30).

A, A. Capsule of kidney.
B, B. Pyramids of Ferrein.
C, C. Cortical labyrinths.
D, D. Malpighian bodies. Many of the glomeruli drop out in the course of preparation, and such empty capsules of Bowman appear as light circular spots.
E, E. Interlobular arteries.
F, F. Boundary region.
G, G. Transverse sections of vessels of the arcades.
H. Base of a Malpighian pyramid.

HUMAN KIDNEY. SECTION PARALLEL WITH MALPIGHIAN PYRAMID. STAINED WITH HÆMATOXYLIN AND EOSIN

OBSERVE: (Fig. 118)

(**Naked eye.**)

1. The **thickness of the cortex**, and its **granular appearance** as compared with the medullary portion.

2. The **markings of the cortex**. (These consist of alternating light and dark lines, radiating from the bases of the Malpighian pyramids. The lighter masses consist largely of collecting tubes, together with ascending limbs of Henle's looped tubes—otherwise called **medullary rays**. Between these lighter areas the dark **labyrinths** appear, in which, by careful attention, the Malpighian bodies may be made out as minute red dots.)

3. A region just outside the medullary pyramids—not as well marked as the outer cortex, in which few Malpighian bodies are seen—the **boundary region**.

4. The finely striated medullary or **Malpighian pyramids**. (The section will usually include portions of two of the last.)

5. That the **bases of the pyramids** do not appear as a sharply defined line, but fade into the boundary region; while the union of the latter with the cortex proper is equally ill defined.

(**L.**) Fig. 118.

1. The **cortical labyrinths**, in which search for—

    a. Portions of the **interlobular arteries**, together with the smaller twigs of the **arterial arcade**.

    b. The **Malpighian bodies**. (The tuft or glomerulus which, with this power, appears as a granular mass, is wanting in numerous places, as indicated by the empty capsules.)

    c. The remaining area occupied largely by the **convoluted tubes**, proximal and distal.

2. The **pyramids of Ferrein**. (Observe that, as they pass into the pyramids of Malpighi, they are well defined, but that they are lost as they approach the region of the capsule.)

(**H.**) Fig. 119.

1. **Malpighian body.** (Select, after searching several fields, a specimen which shows either the **afferent** or **efferent vessel** of the glomerulus. It will be very difficult to find a **capsule connected with the neck of a proximal convoluted tube**, as it rarely

happens to be so sectioned. One may indeed be obliged to examine a dozen slides before succeeding.) Note—

a. The **capsule** (of Bowman or of Müller). (Observe its thickness, as this becomes important in connection with the pathology of the kidney.)

b. The **flattened cells lining the capsule.** (Many of them will have become detached in the preparation of the section.)

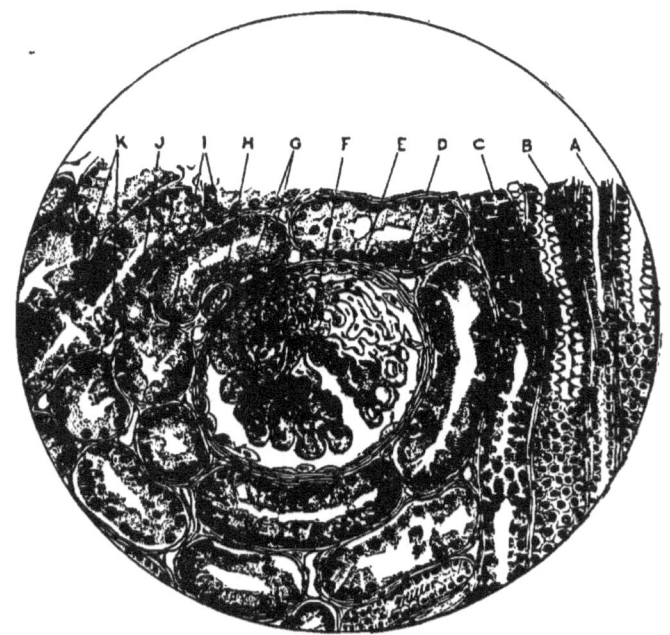

FIG. 119. PART OF THE CORTEX OF HUMAN KIDNEY. HIGH-POWER. SAME SPECIMEN AS FIG. 118 ($\times$ 400).

A. Ascending limb of Henle's loop.
B. Collecting tubule—longitudinal section.
C. Collecting tubule. The upper part of the tube is not sectioned, but shows the attached bases of the lining cells, and thus simulates pavement epithelium. A, B, and C are in a pyramid of Ferrein.
D. Capsule of a Malpighian body. The emerging tubule is not shown, as the body is in T. S.
E. Flattened lining cells of D.
F. Glomerulus.
G. Efferent arterioles.
H. Afferent arteriole.
I. Convoluted tubules.
J. Ascending limb of Henle's loop.
K. Intertubular capillaries.

c. The **glomerulus.** (The great number of nuclei obscures the loops of capillaries. Remember that the nuclei belong partly to the vessels and partly to the flattened cells covering the glomerulus. Endeavor to find **transversely divided loops of the vessels,** showing blood within.)

d. That the **glomerulus does not entirely fill the capsule.**

e. That the **glomerulus is frequently divided.**

f. That the **glomerulus is usually in contact with the capsule at some one point,** where search may be made for

g. The **afferent and efferent arterioles.** (The afferent is more frequently demonstrable, and may be differentiated by its large size, and the connection, which can often be traced, with the interlobular artery.)

2. **Convoluted tubules.** (The convoluted tubules found just beneath the capsule of the kidney generally belong to the distal variety, and they are not as favorable specimens as the deeper proximal portions, on account of the crowding of the tubular elements in the outer cortical regions. Select a transverse section and observe:)

a. The thin **membrana propria,** or wall of the tube. (It does not appear to be made up of fibrillated connective tissue, but has a rather homogeneous structure. Nuclei, however, may occasionally be seen, which apparently belong to this tissue.)

b. The **peculiar lining cells.** (They are unlike any other parenchymatous elements found in the body. Note that, while they are evidently of the columnar or cylindrical type, they differ greatly in form and size. The protoplasm is hazy, granular, and frequently striated. They take a dirty brick-red hue from the eosin.)

c. **The lumen.** (Compared with the diameter of the tube-wall, the lumen is very small, and presents a stellate figure. The urine, in passing through the tubule, is, consequently, brought in contact with a very considerable portion of the lining.)

3. The **large proportion of the cortical area occupied by the convoluted tubules,** and the exceedingly **small amount of intertubular connective tissue.**

4. The **intertubular capillaries.** (These are exceedingly small and difficult of demonstration unless filled with blood. The nuclei of the endothelial wall are frequently seen. The cells of the convoluted tubules are not infrequently detached from the membrana propria, and the space so formed may be mistaken by the careless observer for longitudinal sections of capillaries. These vessels are much better seen in an injected kidney; although if an organ be

selected containing considerable blood, and the corpuscular elements have their color preserved [as in bichromate hardening], the vessels will be easily demonstrated.)

5. **Ascending limb of Henle's loops** in the cortical labyrinths. (The general course of these tubules is confined to the pyramids of Malpighi and Ferrein; but occasionally one of them may be seen passing in a tortuous course toward the outer cortex, running between the proximal convoluted elements. They are easily recognized by their small size and relatively large lumen. They are lined with short columnar or cuboid cells, which stain deeply blue with the hæmatoxylin.)

6. **The pyramids of Ferrein.**

a. **Collecting tubes.** (These will generally be recognized by their large size and the blue color of the staining. They are lined with columnar cells, which are hexagonal in transverse section; and this gives an appearance like pavement epithelium when they are seen from above or below. Endeavor to find a tube split through the center longitudinally, and note the typical columnar cells, as they project inward from the membrana propria toward the now open lumen.)

b. **The spiral tubules.** (These resemble somewhat the convoluted tubules, especially as their cells take much the same dirty red color. The cells, however, plainly columnar, are large and hexagonal in transverse section. The lumen is small.)

c. **Ascending limbs of Henle's loop.** (These are small tubes, and have already been described.)

d. **The intertubular capillaries.** (Inasmuch as, in the specimen under consideration, the vessels of the pyramids are mostly in tranverse section, they are not readily made out, especially if the blood-corpuscles have become decolorized.

7. **Elements of the medullary portion.** Fig. 120.

a. **Collecting tubes.** (These tubes constitute a large proportion of the medulla of the organ. They have already been described. As the apex of the Malpighian pyramid is approached, and the straight tubules unite to form the principal collecting tubes, these again uniting to form the papillary ducts, the lining cells will be seen to get shorter and the lumina larger.)

b **Spiral tubes.** (These can, in many instances, be followed down from the pyramids of Ferrein; and examples are frequently seen very near the pelvis of the kidney in the cortical columns.)

c. **Descending limbs of Henle's loop.** (These tubes are the most difficult of all the tubuli uriniferi to demonstrate. The section must be very thin, and even then they may be mistaken for blood-capillaries. Their peculiar feature consists in the wavy lumen, which is produced by the alternate disposition of the lining cells.)

d. **Loops of Henle.** (The loops will be recognized by the curving of the tube. They are lined with short columnar cells,

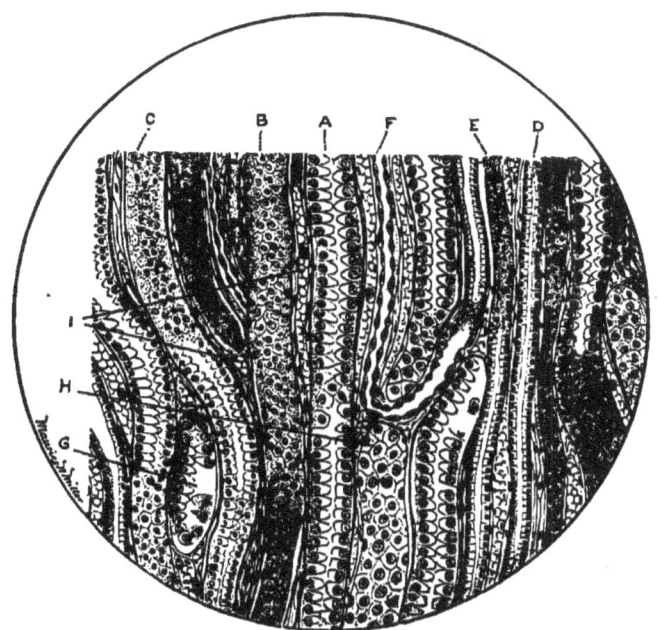

FIG. 120. MEDULLARY PORTION OF SPECIMEN SHOWN IN FIG. 118 (×400).

A. Collecting tubule in L. S.
B. Collecting tubule from above, showing attached bases of lining cells.
C. Collecting tubule presenting different appearance of lining cells, according to mode of section.
D. Ascending limb of Henle's loop.
E. Same as last. The upper end of the tubule not sectioned.
F. Descending limb of Henle's loop. Below may be seen the loop and ascending limb.
G. Oblique section of large collecting tubule.
H. Basal attached extremities of cells lining a large collecting tubule.
I. Intertubular capillaries.

which are sharply brought out by the hæmatoxylin. On account of their course, but few complete sections are seen.)

e. **Ascending limbs.** (Conveniently traced from the loops.)

f. **Intertubular blood-vessels.** (Do not mistake tubules containing blood for capillaries. The human kidney is rarely absolutely normal; and blood is frequently found outside the proper channels. The vessels will be differentiated by the histology of their walls. Quite a number of venules will be seen running in groups in the medulla—the *venulæ rectæ*.

8. **The same elements as in 7** (shown in a transverse section of the middle of a Malpighian pyramid, Fig. 121).

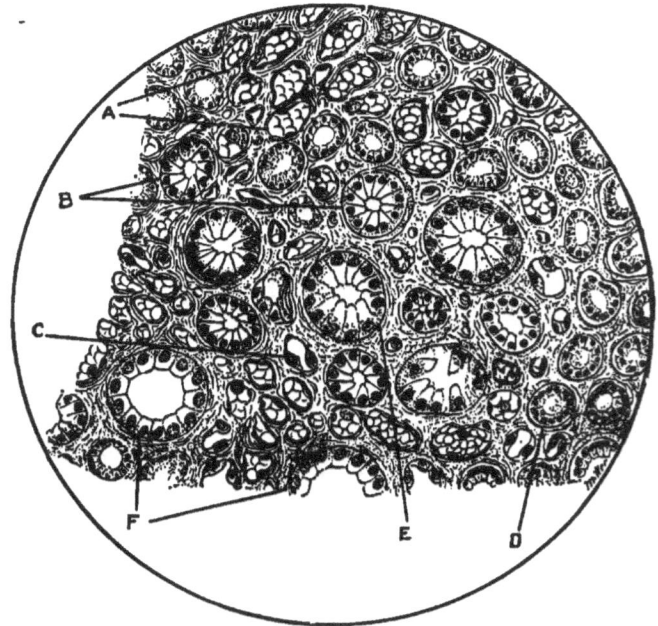

FIG. 121. TRANSVERSE SECTION OF PYRAMID OF MALPIGHI. SAME TISSUE AS SHOWN IN FIG. 118. STAINED WITH HÆMATOXYLIN AND EOSIN (×400).

A. Group of intertubular blood-vessels.
B. Collecting—straight—tubules.
C. Descending limb of Henle's loop.
D. Ascending limb of Henle's loop.
E. Principal—collecting—tubule.
F. Principal tubule. Lower portion near the papillary duct.

The ring of cells will be seen detached from the membrana propria in some instances. This is due to contraction of the tissue during the hardening.

In amphibians the epithelial cells at the neck of the capsule of Bowman are ciliated. In man and the mammals the epithelial cells of the convoluted, spiral, and irregular tubules and the ascending limb of Henle's loop have distinct, vertical striations close to the basement membrane. The cells of the convoluted tubules show variations in size and in the number of granules, dependent on the state of secretion.

DIAGRAM SHOWING DISTRIBUTION OF TUBULES (AFTER PIERSOL)

CORTEX.

Labyrinth.

Malpighian bodies.
Necks of the tubules.
Proximal convoluted tubules.
Irregular tubules.
Distal convoluted tubules.
Collecting tubules.

Medullary Ray.

Spiral tubules.
Ascending limbs of loops of Henle.
Collecting tubules.

Loops of Henle.
Descending limbs.
Ascending limbs.
Collecting tubules. } MEDULLA.

PELVIS OF THE KIDNEY AND URETER

The coats of the ureter are three in number—mucous, muscular, and fibrous. The muscle is unstriated, and is divided into inner longitudinal and outer circular layers. The pelvis of the kidney, and the calyces and infundibula present a similar structure. The circular muscular fibers are numerous around the papillæ, and form a kind of sphincter.

Obtain the ureter of a human subject, if possible. Fix the fresh ureter with alcohol or Müller's fluid. Imbed in celloidin or paraffin. Cut thin sections, stain with hæmatoxylin and eosin, and mount as usual.

OBSERVE:
(L.)
1. The **relative thickness of the epithelium.**
2. The **narrow mucosa.**
3. The **internal longitudinal muscular layer.**
4. The bundles of the **external circular muscular layer.**
5. The **arteries** between the muscular bundles.
6. **Adipose tissue,** more or less abundant in the loose cellular tissue surrounding these canals. (This element will afford a prominent feature of the section of the pelvis of the kidney, while the muscular tissue will be seen to a limited extent only.)
(H.) Fig. 122.
7. **The epithelium.** (*a*) That it is of the thin, **stratified squamous** type known as **transitional.** (*b*) The broad basal attach-

M

ment of the deep cells. (c) The elongated form of the cells generally. (d) The more flattened surface cells. (e) The outline of the last, as seen in the detached specimens. (f) The deeper cells present tapering prolongations, generally at one end only,

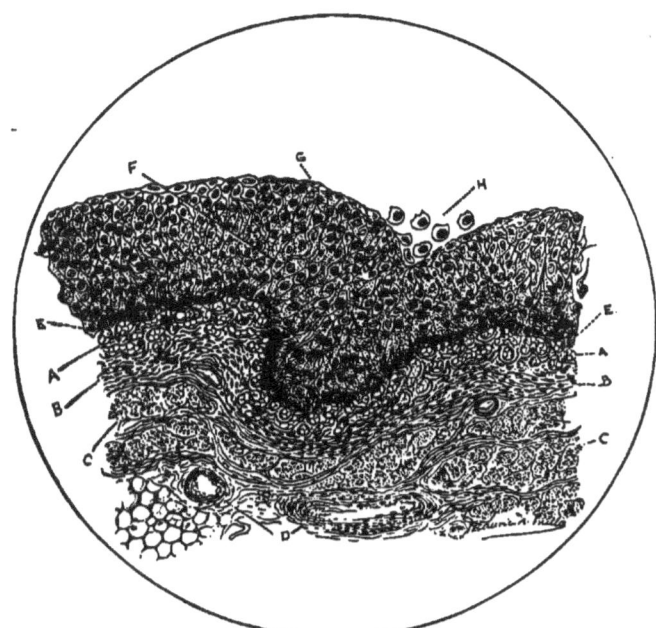

Fig. 122. Section of the Ureter near the Pelvis of the Kidney. Stained with Hæmatoxylin and Eosin (×400).

A. Rich capillary plexus of the mucosa.
B. Internal muscular coat.
C. External muscular coat.
D. Large vessels of the areolar adventitia.
E. Deep layer of somewhat cubical cells.
F. "Tailed cells" of the middle epithelial layers.
G. Surface cells in profile.
H. Detached surface cells.

and are hence called "tailed cells." They may be confounded with similarly shaped cells from the bladder.

THE URINARY BLADDER.

The layers of the bladder are mucous, muscular, and fibrous, while part of it has a serous covering derived from the peritoneum. The muscle is unstriated, and is arranged in inner and outer longitudinal layers, with a circular layer between them. The muscular layers are not well marked. The internal vesical sphincter is formed from an increase in the thickness of the circular layer. Numerous minute ganglia are found along the nerves of the bladder,

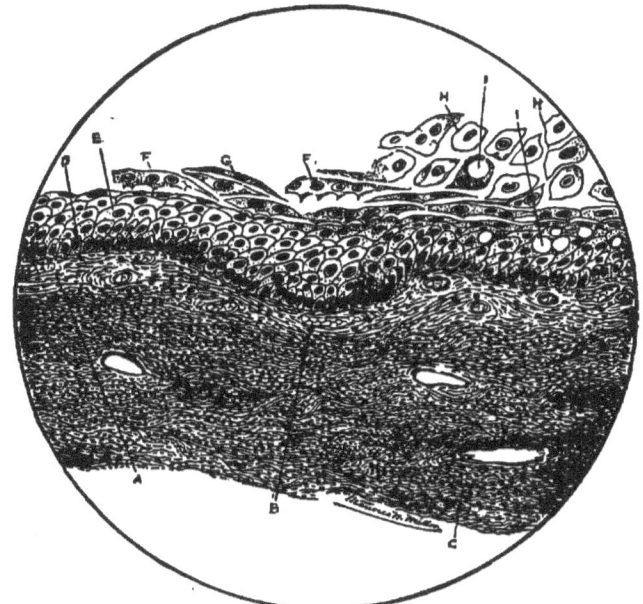

Fig. 123. Vertical Section of the Lining Portion of the Bladder (Male) behind the Trigone. Stained with Hæmatoxylin and Eosin (×400).

    A. Connective tissue of sub-epithelial region.
    B. Capillary supply of sub-epithelial region.
    C. Muscular wall of bladder.
    D. Deep cells of the epithelial lining.
    E. Middle region of lining.
    F. Detached surface cells, showing processes beneath.
    G. Thin surface cells in profile.
    H. Squamous surface cells, seen detached, in plan.
    I. Vacuolated cells.

which contain both medullated and non-medullated fibers. As in the ureter, the blood supply of the mucous and muscular coats is abundant. Small nodules of lymphoid tissue occur in the mucous membrane. The epithelium is transitional; but the disposition of the cells varies with the extent to which the bladder is expanded.

Prepare sections of human bladder in the same manner as sections were made of the ureter.

OBSERVE :
(L.)
1. The **epithelial lining**. (a) That it is formed after the **transitional type**.
2. The **mucosa** and its **capillary supply**.
3. The **dense muscular portion**.

(H.) Fig. 123.
4. **The epithelium**. (a) The size of the cells. (b) The layers of the epithelium, which are **deep, middle, and superficial**. (c) That the deep cells are polyhedral or columnar. (d) The form of the middle cells; not unlike those of the corresponding region in the ureter. (e) The large, scaly **surface epithelial cells**. (Note that while these all appear flat, when seen in plan, it is only those of the extreme surface that are simple scales; the less superficial examples show, when viewed in profile, **prolongations from the under surface,** by means of which union is effected with the deeper cells.)

THE URETHRA

The urethra consists of a mucous coat surrounded by muscular and fibrous tissue. The muscle is unstriated. Numerous papillæ cover the surface of the mucous membrane proper. The lining in the *female* urethra is stratified squamous epithelium. A few small glands are found in the female urethra. The lining of the *male* urethra varies in the different regions. In the prostatic region it is transitional epithelium, as in the bladder; in the membranous portion it is stratified columnar; in the spongy or penile it is simple columnar; while in the fossa navicularis it becomes stratified squamous epithelium, and is continuous with that covering the glans penis. Small, branched, mucous glands (the glands of Littré) occur throughout the male urethra.

THE FEMALE GENERATIVE ORGANS

THE VAGINA AND UTERUS

The walls of the vagina are lined with a *mucous membrane*, covered with a thick, stratified squamous epithelium, beneath which are numerous papillæ. The *submucous*, unstriated *muscular*, and fibrous *adventitious* coats are not well defined. Glands are not present.

The uterus has *mucous*, *muscular*, and *serous* layers, but the muscular layer is much the thickest. The muscle is unstriated. The fibers increase both in size and number during pregnancy. The mucous membrane is covered with a single layer of columnar, ciliated epithelial cells, which create a current in the direction of the cervix. Numerous tubular glands, lined with similar cells, extend down to the muscular layer. During menstruation the mucous membrane, having become thickened, soft, and distended with blood, undergoes disintegration in its outer portion, which is cast off. Accompanying this phenomenon there is an escape of blood from the capillaries. The epithelium is quickly renewed from the deeper portions of the glands of the mucous membrane.

PRACTICAL DEMONSTRATION

From the body of a (preferably young) human female, as soon as possible after death, remove centimeter cubes of the organs required, observing that the lining is included. The outer portions are of very little moment comparatively. Secure pieces from the os, cervix, and fundus of the uterus, and the wall of the vagina.

The vagina and uterus of a human subject should be secured, because their structure differs considerably from that of the same organs in the lower animals. The differences are greater than is the case with many of the organs hitherto studied.

We desire to prepare the tissue so as to keep the original form of cell-elements—to avoid contraction—and Müller's fluid will accomplish this perfectly. Allow the pieces to remain for a month in the bichromate solution, with an occasional change. Complete the hardening in alcohol, after washing as usual. Infiltrate with celloidin, and let the sections be vertical to the mucous surface. The tissues should not be handled with the fingers; otherwise the epithelial lining cells will be detached. Stain with hæmatoxylin and eosin; mount in balsam.

VAGINA AND UTERUS OF THE HUMAN FEMALE AT PUBERTY—VERTICAL DEXTRO-SINISTRAL SECTION OF THE RIGHT LIP OF THE OS, AND INCLUDING PART OF THE VAGINAL CUL-DE-SAC

OBSERVE:
(L.)
1. The outline of the section. (Commencing at D, Fig. 124, which is placed in the internal os, follow downward, out upon

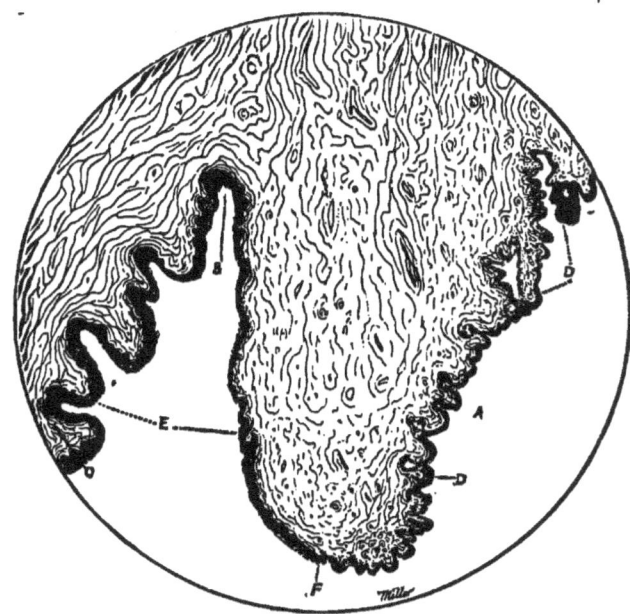

FIG. 124. VERTICAL DEXTRO-SINISTRAL SECTION OF THE RIGHT-HAND SIDE OF THE OS UTERI. SHOWING THE INTERNAL OS, THE EXTERNAL OS, THE VAGINAL CUL-DE-SAC, AND THE UPPER PORTION OF THE VAGINAL WALL (× 60).

    A. The letter is placed in the internal os.
    B. Vaginal cul-de-sac.
    C. Vaginal wall.
    D. Columnar epithelium of the internal os. In the upper portion the tubular glands are well seen.
    E. Stratified epithelium of the vaginal lining.
    F. Change at the external os from stratified flattened to columnar epithelium.

the external os, curve upward, reaching, at B, the vaginal cul-de-sac. Descend along the right vaginal wall.)

2. The irregular surface of the internal uterine wall. (Caused by longitudinal section of the glandulæ uterinæ or glan-

dulæ utriculares—branched tubular glands. These are increased in depth during pregnancy, and are most prominent in the lower portion of the organ.)

3. **The epithelium.** (*a*) The **deeply stained layer, lining of the vagina, cul-de-sac,** and **external os.** (*b*) The **wavy course of** *a* as it covers the irregularly formed and often imperfect **papillae of the mucosa.** (*c*) The lighter appearance of the lining of the internal os. (*d*) **Projection of the last into the glands.** (*e*) The **sharp line of separation between** the deeply stained lining common to the **vagina and** the lighter lining of the uterus at the **external os** (Fig. 124, F).

4. The **mucosa of the uterus.** (There are no sharply defined regions in the genito-urinary tract corresponding to the mucosa and submucosa of typical mucous membranes. The arrangement generally is, (1) an epithelial lining; (2) a subepithelial structure, consisting of a more or less prominent or abundant plexus of capillaries supported by delicate connective tissue, and which corresponds to the typical **mucosa;** (3) the muscular walls proper, consisting of layers in different directions, frequently irregularly disposed and seldom in distinct fasciculi.)

5. The **mucosa of the vagina** (the surface of which is beset with small papillæ, and in which large veins are prominent).

6. The **uterine and vaginal walls** (consisting largely of involuntary muscular fibrils, recognized by the elongated and deeply stained nuclei, and containing numerous thick-walled arteries and irregular lymph-spaces.)

(H.)

7. The **uterine epithelium** (Fig. 125). (*a*) That it consists of a **single layer** of cells. (*b*) That the **cells are columnar or cylindrical.** (*c*) The cells in transverse section are **polygonal.** (*d*) They are **ciliated.** (If the section has been properly prepared from uninjured tissue, the cilia will be seen without difficulty, and especially in the depressions where they are somewhat protected.) (*e*) The **cell-body** and **nucleus.** (Note the elongated, clear, free portion, and the frequent curving of the whole. Near the attached extremities, which often appear pointed, note the small, deeply stained nuclei.) (*f*) The **large mucous** crypts which occur in the cervix. Retention of their secretion produces the cysts which are known as the **ovula Nabothi.** (*g*) It is difficult to see the **basement membrane.**

8. The abrupt **transition from columnar to flattened cells** in the epithelium of the external os. (*a*) The **shortening of the columnar cells** as the point of change is approached. (Sections must be examined until one is found showing this point well. The illustration [Fig. 125] is not exaggerated, and a properly cut and selected specimen must exhibit clearly the last columnar and the adjoining flattened cell.)

9. The **vaginal epithelium** (Figs. 125 and 126). (*a*) That it is

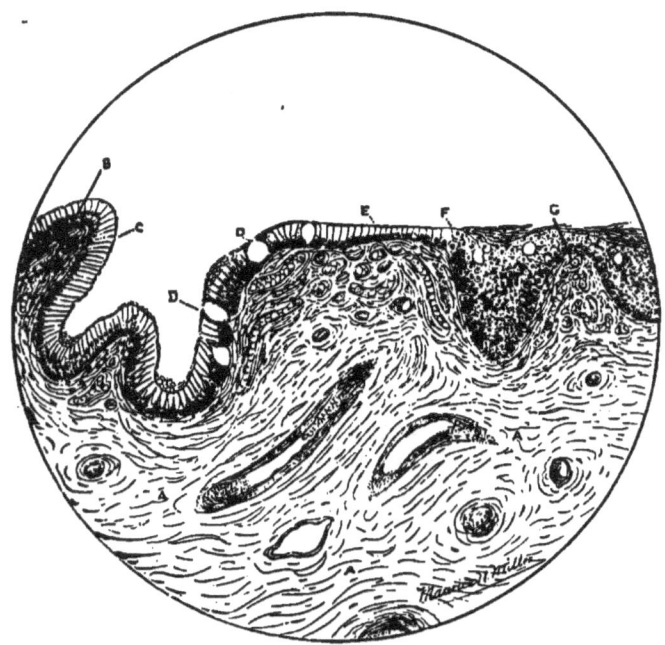

Fig. 125. External Os of Fig. 124. More Highly Magnified (×400).

A. Muscular tissue of the os uteri, with numerous blood-vessels.
B. Capillary plexuses of subepithelial tissue—mucosa.
C. Ciliated columnar cells covering the os.
D. Vacuolated cells.
E. Shortening of the columnar cells preparatory to
F. Change from typical uterine epithelium—ciliated columnar cells—to flattened stratified cells.
G. Papillary structure of the mucosa of the external os, after change of epithelium.

of the **stratified** squamous variety. (*b*) The **deepest line of cells** following the sinuous line formed by sectioning the papillary mucosa. (*c*) That the **cells** are more or less **flattened**. (*d*) That their edges, excepting those of the surface, are **serrated**. (*e*) The change in form as the surface is approached. (*f*) The **surface cells.** (These are very much flattened, and so fused in longitudinal section as to resemble fibers.) (*g*) **Detached surface cells.** (At H, Fig. 126, these are shown in plan, having been torn off; those intact are, of course, seen in profile.) (*h*) The **nuclei,** evenly granular, usually larger than a red blood-corpuscle.

10. The **subepithelial vaginal structures.** (*a*) The large and abundant **capillaries** of the mucosa. (*b*) The **submucosa,** not

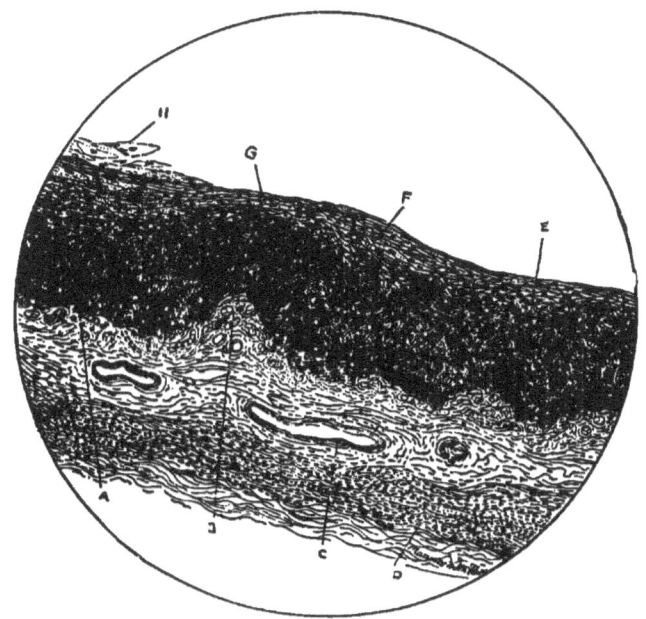

FIG. 126. VERTICAL SECTION OF THE VAGINAL LINING AT PUBERTY. STAINED WITH HÆMATOXYLIN AND EOSIN (× 400).

A. Subepithelial capillary plexus.
B. Papillary arrangement of the mucosa.
C. Large blood-vessels in the submucosa.
D. Muscular wall of vagina.
E. Deep cells of the lining epithelium.
F. Middle strata of lining stellate cells.
G. Surface cells in profile.
H. Surface cells in plan—detached.

clearly separated from the mucosa, but easily recognized by the large vessels and the abundant connective tissue. (*c*) The muscular portion of the vaginal wall.

THE FALLOPIAN TUBE OR OVIDUCT

The Fallopian tube has a *serous coat* derived from the peritoneum; a *muscular layer*, consisting of unstriated muscular fibers, mostly disposed in a circular manner; and a *mucous membrane* which lines the tube. The mucous membrane is covered with a

Fig. 127. Fallopian Tube. Transverse Section of a Fold of the Ampulla—after Henle.

* * Spaces between the folds.

single layer of columnar, ciliated epithelial cells, whose cilia create a current in the direction of the uterus. The mucous membrane has numerous longitudinal folds, which are very complicated, and which, in tranverse section, appear like glands. Glands, however, are not present.

PRACTICAL DEMONSTRATION

Harden the Fallopian tube of a healthy young female in alcohol or Müller's fluid; imbed in celloidin; cut transverse sections; stain in hæmatoxylin and eosin; mount in balsam.

(L.)

Find the three principal layers: the **mucous membrane**, with its folds; the **muscular layer**, consisting of unstriated muscle, divided into an inner circular and an outer longitudinal band; and the thin **serous** covering.

(H.)

The mucous membrane consists of a fibro-elastic basis, covered by a single layer of **columnar ciliated epithelial cells**, which will be found well preserved at the bottoms of the folds.

THE OVARY

The ovary consists of a stroma or ground-substance of connective and smooth muscular tissue, in which are scattered various sized spherical bodies, the *Graafian follicles*.

The stroma is divided into three layers or regions, which are not very sharply differentiated.

The free surface of the ovary is covered with a single layer of low columnar epithelial cells, called the *germinal epithelium*.

Immediately beneath the epithelium is a thin layer of fibrous tissue, termed the *tunica albuginea*.

The *cortex* proper, or second layer, is distinguished by the Graafian follicles, which will be described later.

The central portion of the organ, the *zona vasculosa*, is largely occupied by thick-walled blood-vessels, among which the extremely tortuous arteries are especially evident. Occasionally one may see in this region somewhat ovoid nodules in varying degrees of retrograde change—the *corpora lutea*.. They present the phenomena resulting from the maturation of the follicle during menstruation. The accompanying illustration, Fig. 128, was drawn from a corpus luteum which had formed in the site of a Graafian follicle, the contents of which had escaped at some menstrual epoch, and been *followed by impregnation*.

PRACTICAL DEMONSTRATION

The ovary of a young animal is to be preferred. If the organ cannot be obtained from the human subject, the female of almost any domestic animal will provide an excellent demonstration for the histological elements. Let the tissue be hardened with strong alcohol, and sections be cut vertically to the free surface and stained with hæmatoxylin and eosin. The sections should include at least one-half the depth of the organ, so as to exhibit all of the regions.

SECTION OF THE ADULT HUMAN OVARY

OBSERVE: (Fig. 128)

(L.)

1. The **tunica albuginea.** (Note that the layer is not of uniform thickness, and is composed largely of spindle-shaped cells, as

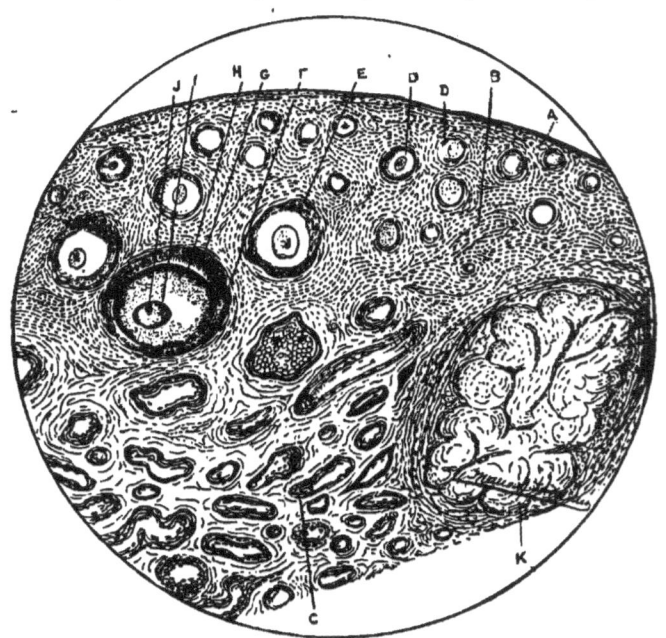

FIG. 128. SECTION OF AN OVARY FROM A WOMAN 35 YEARS OLD. STAINED WITH HÆMATOXYLIN AND EOSIN (×250).

A. Surface of the ovary.
B. Stroma.
C. Large, tortuous, thick-walled arteries of the central portion of the organ.
D, D. Small Graafian follicles of the superficial zone.
E. Larger follicles of the deeper portion.
F. Membrana propria of a Graafian follicle.
G. Membrana granulosa of the follicle. The line leads to the discus proligerus.
H. An ovum.
I. Germinal vesicle.
J. Germinal spot.
K. An old corpus luteum.

shown by the numerous elongated nuclei. Search particularly for and note the character of the epithelial covering.)

2. The **cortical layer,** containing numerous Graafian follicles, and possibly a corpus luteum. (Note the aggregation of the smaller follicles in the extreme outer portion of the region.)

3. The **zona vasculosa**. (Note the unusual thickness of the vascular walls and the irregular outlines on section, on account of their tortuous course.)
(H.)

4. The **Graafian follicles**. (*a*) **Their diameter**, varying from 25 μ when young to 5 or 10 mm. when mature. (*b*) The **membrana propria**. (This is difficult to separate from the stroma of the ovary itself, except in more mature follicles than shown in the section. (*c*) The **membrana granulosa**. (This, in general, appears to be the outer layer of the follicle, on account of the difficulty of separating the membrana propria from the stroma proper of the ovary. Note that it is composed, in the smaller and less mature follicles, of pavement cells, and that the cells become thicker with maturation, until columnar cells in a single layer result. (*d*) **The ova**. (These are contained within the follicles, excepting that they may have become detached during manipulation of the section, and occupy the greater part of the follicles.) (*e*) The **zona pellucida** (the thin wall of the ovum). (*f*) The **discus proligerus**. (This will be recognized as a mass of polyhedral cells, connecting the ovum at one side with the columnar cells of the membrana granulosa. These cells will proliferate later in the development, and completely enclose the ovum.) (*g*) The **germinal vesicle**. (Contained within the ovum. The contents appear granular; it, as well as the ovum, is fibrillated; but this demonstration cannot be made, excepting the animal be killed for the purpose, and the tissue elements fixed, before changes, which quickly follow death, occur.) (*h*) The **germinal spot**. (Appearing as a small dot within the germinal vesicle. The ovum presents the characteristics of what it indeed is—a typical cell, with *wall, body, nucleus, and nucleolus*.)

5. The **corpus luteum**. (The example shown in the drawing was developed after the contents of the Graafian follicle, which it represents, had suffered impregnation; and it has arrived at the later stage of the series of the phenomena connected with its development—the stage of cicatrization. The cicatricial tissue, to which the letter K points, indicates the remains of the membrana granulosa. Outside is seen the thickened membrana propria, while among the contents will be found pigment-granules and fat-globules, imbedded in a structureless, gelatinous stroma. This material results from changes in the clot of blood effused after the escape of the ovum.)

FORMATION OF THE OVUM

As has been previously shown, the ovary is covered with columnar epithelium; and, singular as it may appear, the fifty thousand Graafian follicles, which it is estimated are developed during the life of the human female, have their origin in these cells.

During fœtal life this surface epithelium undergoes a very rapid-proliferation, and chains of cells become imbedded in the stroma of the ovary. These epithelial prolongations are called ovarial or *egg-tubes*. A little later in the development, separate portions or links of these chains are cut off by the ingrowth of the stroma. The little groups of cells thus isolated become each a Graafian follicle.

Scattered among the columnar cells, larger, more nearly spherical cells are also found, the *primordial ova*. These are also imbedded in the substance, and one at least will always be found among the minute collections of cells which have been isolated.

In the process of development, each group of cells becomes a Graafian follicle with its contained ovum, the columnar cells forming the wall proper, and the primordial cell the ovum, with its vesicle and germinal spot.

PRACTICAL DEMONSTRATION

The ovary from a still-born babe is to be removed with the scissors, exercising the utmost care that the surface be not touched. The ovary of a rabbit or guinea-pig may also be used with advantage. Place immediately in strong alcohol, and in twenty-four hours it will be fit for cutting. Cut extremely thin sections at a right angle to the free surface and including the same; stain with hæmatoxylin and eosin; mount in balsam.

OBSERVE :
OVARY OF HUMAN INFANT (Fig. 129)

(L.)

1. **The free surface.** (Note the occasional depressions which mark the involution of the epithelial cells.)

2. **The layers.** (Note the absence of demonstrable tunica albuginea and the great area occupied by the cortex. The vessels of the central portions are, unlike the ovary of mature life, large, not numerous, and thin-walled.)

(H.)

3. The primordial ova of the surface epithelium.
4. The projecting lines or chains of epithelium undivided. (Here the cells seem rather elongated.)
5. Chains which are in process of subdivision.
6. Young Graafian follicles in columns at a right angle to the surface of the ovary.
7. The discus proligerus, in many instances still composed of flattened cells.
8. Follicles showing discus proligerus as columnar cells.
9. Follicles showing great proliferation of discus proligerus.

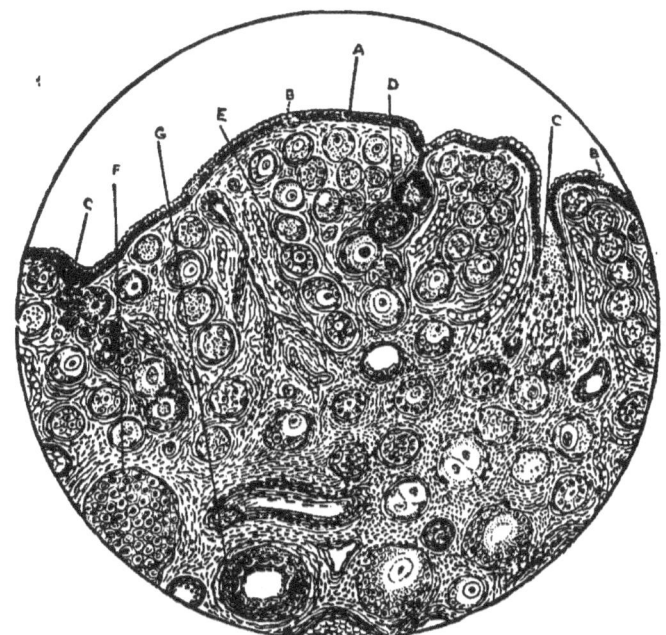

FIG. 129. SECTION OF OVARY OF CHILD. DEATH TEN DAYS AFTER BIRTH (×350).

A. Germinal epithelium, covering surface of the ovary.
B. Primitive ova.
C, C. Projection of surface epithelium within the organ.
D. Constriction of the projected chain or cord of epithelium and isolation of portions to form Graafian follicles.
E. Chain of Graafian follicles. The stroma is seen filled with previously formed follicles which have now become isolated.
F. A large Graafian follicle. It has been cut in half; the ovum has fallen out, and the membrana granulosa is seen lining the cup-shaped cavity.
G. Large arteries of the central portion of the ovary.

208 STUDENTS HISTOLOGY

10. Ova in the early stages of development from primordial cells, with granular vesicles.

11. Instances of development of two, possibly three, ova in a single follicle.

12. Large blood-capillary supply of cortex,—vessels generally parallel with the chains of follicles.

THE MAMMARY GLAND

The mammary gland consists of acini lined by epithelial cells. The acini are grouped into lobules and lobes. There is an abun-

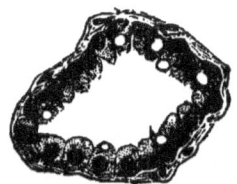

FIG. 130. MAMMARY GLAND OF THE DOG. TRANVERSE SECTION OF AN ACINUS IN THE EARLY STAGES OF FAT FORMATION. (HEIDENHAIN.)

dant framework of fibrous and adipose tissue. The ducts are lined by columnar epithelium, and open at the nipple. The lobules and acini of the resting gland are small. During activity the acini are lined by low cubical epithelial cells resting on a basement mem-

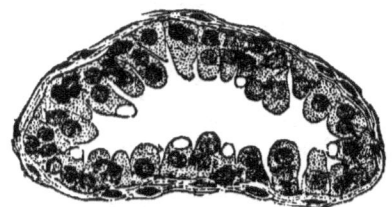

FIG. 131. MAMMARY GLAND OF THE DOG AT THE HEIGHT OF ACTIVITY. (HEIDENHAIN.)

brane. Globules of oil form within these cells, and are discharged, making the oil-globules of the milk.

The secretion of the gland during the first few days of activity contains numerous cells in which drops of oil are present, the **colostrum-corpuscles**. Sections of the resting and active mammary gland of the dog may be studied with advantage.

THE MALE GENERATIVE ORGANS

THE TESTICLE

The testicle is a glandular body (Fig. 132, T), oval in shape and flattened laterally. The epididymis, E, is an elongated and arched structure, which is applied to the upper end, and posterior border of the body of the testicle. It is enlarged at each end. The upper end is termed the *globus major;* the lower, which is somewhat smaller, the *globus minor;* while between them is the *body of the epididymis.*

From the lower end of the epididymis a hard convoluted tube,

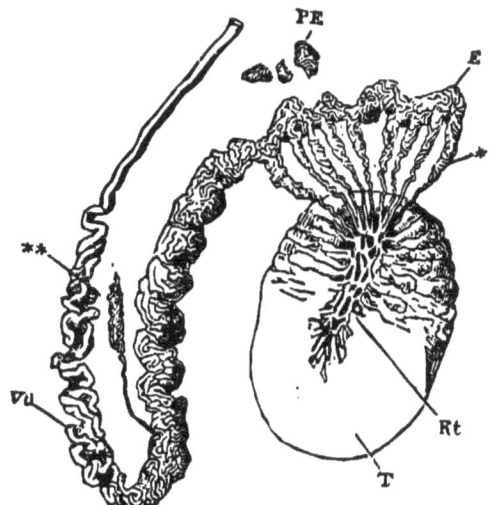

FIG. 132. DIAGRAM OF THE COURSE OF THE CANALS IN THE TESTICLE AND EPIDIDYMIS, AND THE PASSAGE OF THE CANAL INTO THE VAS DEFERENS. (LAUTH.)

T. Testicle.
Rt. Rete testis.
E Epididymis.
PE. Organ of Giraldès.
Vd. Vas deferens.
* Vasa efferentia.
** Vas aberrans.

the *vas deferens,* Vd, is given off, which soon straightens, and extends upward in the spermatic cord.

The testicle is almost completely covered with a serous membrane, the *tunica vaginalis.* Beneath that is a strong fibrous capsule, the *tunica albuginea,* with partitions or septa converging toward the posterior portion, the *mediastinum testis,* or corpus

Highmori. Between the septa are delicate tubular masses—the *lobules*.

The tunica albuginea is the supporting framework of the testicle. Its inner layers convey the blood-vessels, and constitute the *tunica vasculosa*.

The mediastinum receives the blood-vessels and contains also the *rete testis*, Rt, the termination of the secreting tubules.

Two or more seminiferous tubules are contained in each lobule. They are fine thread-like tubes closely packed together—the convoluted portion. As these tubules approach the mediastinum they

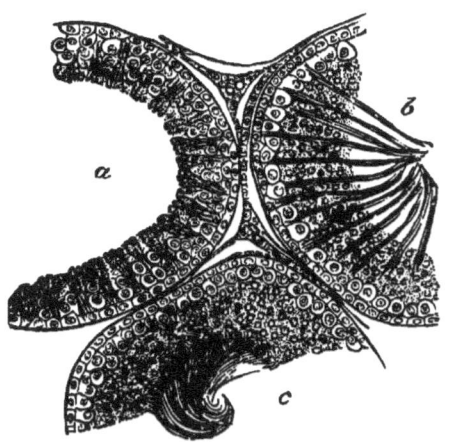

FIG. 133. TRANSVERSE SECTION OF THREE SEMINIFEROUS TUBULES OF THE CAT. (SCHÄFER.)

A. Containing spermatozoa least advanced in development.
B. More advanced.
C. Containing fully developed spermatozoa.

join to form a smaller number of *straight tubes*, or *tubuli recti*. The straight tubules enter the mediastinum, making a plexus of tubular spaces, the *rete testis*, and leave the rete testis at its upper end as fifteen to twenty delicate ducts, the *vasa efferentia*, Fig. 132.

The vasa efferentia, entering the globus major, become coiled into small conical masses, the *coni vasculosi*, and open into a single convoluted tube, the *canal of the epididymis*. This canal is remarkably tortuous; it passes down the back of the body of the testicle into the globus minor, and finally terminates in the vas deferens.

(For the character of the lining epithelial cells of the different parts, see page 55.

PRACTICAL DEMONSTRATION

Harden the testicle of some animal, preferably a rat, in Flemming's, Müller's, or Orth's fluid. After washing, complete the hardening in alcohol. Imbed in celloidin, cut transverse sections, stain in hæmatoxylin and eosin, and mount in balsam.

(L.)

Find the **tunica albuginea** covering the testicle, consisting of fibrous tissue; its outer serous layer; the **tunica vaginalis**, and its looser inner layers, containing many blood-vessels; the **tunica**

Fig. 134. Human Spermatozoa, Highly Magnified. (Retzius.)
A. Seen from above.
B. Head seen from the side.
C. Extremity of the tail.

vasculosa. Find the thickening of the tunica albuginea behind, called the **mediastinum**, from which fibrous septa arise, dividing the testicle into compartments. These compartments are seen to contain **seminiferous tubules** 130 μ to 140 μ in diameter, cut at various angles.

(H.)

Examine seminiferous tubules in transverse section. Observe a lining of epithelial cells in several layers, and **spermatozoa** in the central part of the tube, in various stages of formation. The tails of the spermatozoa point toward the center of the tube, and the heads are directed toward the epithelial lining. Among the epithelial cells some may be found that are undergoing karyokinesis.

Careful study of selected specimens has shown that the epithelial cells next the basement membrane are of two sorts: (1) The **sustentacular cells**. (2) The **spermatogenic cells**. The spermatozoa are derived from the spermatogenic cells.

Spermatozoa may be obtained from the fresh testicle or epididymis of some one of the domestic animals. The human spermatozoön is about 50 μ in length. It consists of a short oval **head**, a short middle piece or **body**, and a long **tail**. The tail possesses active vibratile movements.

THE PROSTATE GLAND.

The prostate gland consists of a number of tubular acini, which are imbedded in a muscular and fibrous framework. The muscle is of the unstriated variety, and is very abundant. The tubular glands are lined by columnar epithelium. They open by a number of ducts into the urethra. Certain small, round, concentrically laminated bodies, the *prostatic concretions*, often occur in the ducts and acini, especially in old subjects. The secretion of the prostate gland is thin and nearly transparent, containing granules and epithelial cells.

ERECTILE TISSUE

Erectile tissue is best demonstrated in the penis, preferably of a human infant, cut in transverse section. The two corpora cavernosa and the corpus spongiosum are readily distinguished. The erectile tissue can be studied in the corpora cavernosa.

The connective tissue surrounding the cavernous bodies gives off numerous trabeculæ, forming a framework, enclosing spaces. These spaces are lined by endothelium, and communicate with one another freely. The arteries open into them, and the veins open from them. During erection they become greatly distended with blood. Similar erectile tissue occurs in the external generative organs of the female.

THE SUPRARENAL BODY

This body is attached by areolar tissue to the summit of the kidney, and consists of several folia or leaflets. An examination of one of these leaflets will give us an idea of the organ as a whole. The plan of structure seems to be as follows:

In the connective tissue which supports the folia are found arterial branches derived from the phrenic and renal arteries, besides the suprarenal artery itself. These arteries penetrate the organ, break up immediately into capillaries, which finally converge toward the center of the leaflet; the blood is here collected in thin-walled veins, by which it is drained into the suprarenal vein, thus leaving the body.

The capillary meshes vary in form and size, according to their position. Near the circumference of the leaflets the meshes are small and ovoid; while, as the center is approached, they become elongated. These spaces between the capillaries are filled with compressed, globular, nucleated cells, the smaller containing only perhaps six or eight, while the longer may be occupied by thirty or forty of these cell-elements, which constitute the parenchyma of the organ. This variation in size of the cell-compartments, contributing, as it does, to alter the appearance of the different zones of the tissue, has given rise to a division into *cortex* and *medulla*. The cortex is divided from without inward into a zona glomerulosa, zona fasciculata, and zona reticularis. Many nerve-fibers enter the organ with the arteries. They form a plexus, mostly of non-medullated fibers, in the medulla. Ganglion-cells are numerous. The surface is covered with a fibrous capsule continuous with the supporting framework.

The suprarenal body is also often known as the suprarenal capsule. This body, the spleen, the thyroid and the thymus glands, with certain other organs of less importance, are sometimes called "*the ductless glands.*"

PRACTICAL DEMONSTRATION

The tissue is best hardened in strong alcohol, and should be cut as soon as the hardening is complete. It will be sufficiently firm to admit of the thinnest sections being made free-hand or with a simple microtome. The sections, stained with hæmatoxylin and eosin, give excellent differentiation.

214 STUDENTS HISTOLOGY

HUMAN SUPRARENAL BODY. (Figs. 135 and 136)

SECTION OF A SINGLE LEAFLET, CUT TRANSVERSELY TO THE CENTRAL VEINS. STAINED WITH HÆMATOXYLIN AND EOSIN.

OBSERVE:

(L.)

1. Section of **arterial twigs** on the border of the leaflet.
2. The **convergence of the parenchyma** toward the center.
3. The **large and thin-walled central veins.**
4. The **small size of the parenchymatous areas** on the outer borders and their elongation within, making the zona glomerulosa and the zona fasciculata respectively.
5. Distinguish the **zona reticularis** next to the medulla.

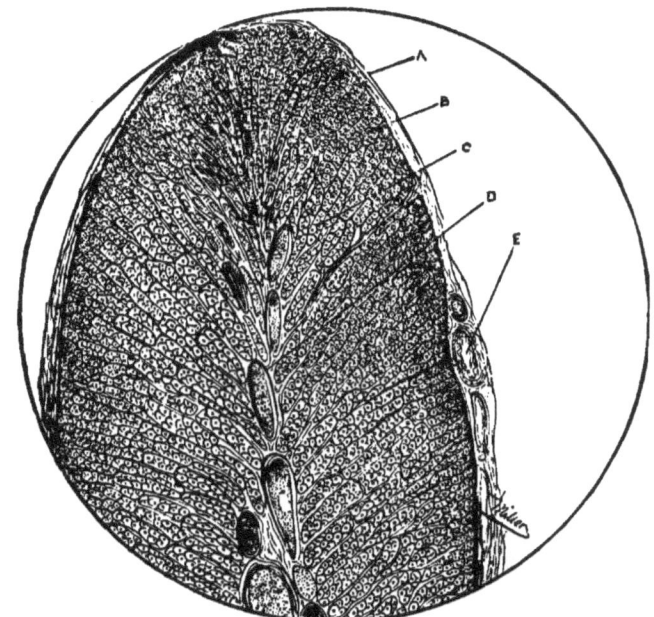

FIG. 135. VERTICAL SECTION OF A SINGLE LEAFLET OF THE SUPRARENAL BODY. STAINED WITH HÆMATOXYLIN AND EOSIN (× 60).

A. Fibrous tissues surrounding and connecting the leaflets.
B. The outer portion, consisting of small compartments—the so-called zona glomerulosa.
C. The central elongated cell-compartments and medulla.
D. Large, thin-walled central veins.
E. Arteries ramifying in the outer fibrous tissue which supply the parenchyma.

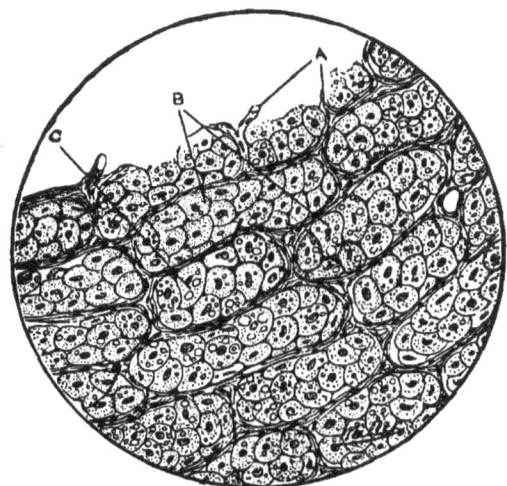

FIG. 136. SAME SECTION AS FIG. 135, MORE HIGHLY MAGNIFIED. ZONA FASCICU-LATA ($\times$ 400).

A. Blood-capillaries, arising from the arteries seen in the preceeding illustration, and ramifying in the connective tissue framework.

B. Compartments—lobules—formed by delicate connective tissue prolongations from the fibrous capsule.

C. Lobular parenchyma. These large, somewhat rounded cells are generally mononucleated, contain fat-globules, and are frequently pigmented.

(H.) Fig. 136.

1. The **capillary plexus**, forming ovate or **elongated meshes**.

2. The compressed **globular cells of the parenchyma**. (Note that the cells are small in the small compartments, as though crowded. This is due, in a measure, to the contraction of the tissue from the rapid hardening.

3. The minute **fat-globules** in the parenchyma.

4. Yellow **pigment-granules** in the cells, especially of the medulla.

THE NERVOUS SYSTEM

Structural Elements

The elements of the nervous system are:
1. Nerve-Fibers.
2. Nerve-Cells.
3. Connective Tissue and Neuroglia.
4. Peripheral Termini.

NERVE-FIBERS

Nerve-fibers are of two sorts, medullated or white, and non-medullated or gray.

A typical *medullated nerve-fiber* consists of three portions; viz., a central conducting portion, the *axis-cylinder;* the medullary sheath, or *white substance of Schwann;* and the enveloping connective tissue substance, the *neurilemma.* Such fibers are found largely in the trunks of the cerebro-spinal system, while medullated fibers devoid of the neurilemma exist in the optic and acoustic nerves, the spinal cord, and the brain.

The axis-cylinder may be seen to be split up longitudinally, and is found to be composed of fine primitive or ultimate fibrillæ, which may present minute varicosities or swellings at irregular intervals.

The white substance of Schwann, which is largely fatty, presents under the microscope the most prominent feature of medullated nerves, affording a nearly complete investment of the nerve-axis.

The neurilemma is an elastic envelope, which completely invests the medullary substance. This tubular membrane is nucleated, and at intervals is constricted so as to reach the axis-cylinder. These constrictions are called the *nodes of Ranvier.* The neurilemma presents a single nucleus, with a small amount of protoplasm between each two of these nodal points. The constrictions do not, however, affect the even caliber or continuity of the axis-cylinder.

The distance between two nodes of Ranvier may be as much as a millimeter, or it may be less than that.

Non-medullated nerve-fibers are found chiefly in the sympathetic system. But some non-medullated fibers occur in most of the cerebro-spinal nerves, and they form the chief part of the olfactory nerve. Non-medullated nerve-fibers have an axis-cylinder and a neurilemma, but no medullary sheath. They branch freely, and aid in the formation of plexuses. There are numerous

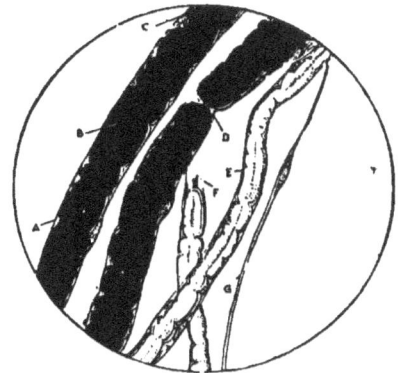

Fig. 137. Separated Nerve-Fibers (× 400).

A. Neurilemma of a fiber.
B. White substance of Schwann, stained with osmic acid, which hides the axis-cylinder.
C. Nucleus of the neurilemma.
D. One of Ranvier's nodes in an osmic acid-stained fiber, showing the axis-cylinder between the separated portions of Schwann's sheath.
E. A medullated fiber, teased in normal salt solution. The medullary substance has become coagulated on exposure and removal. The axis-cylinder is faintly seen.
F. Axis-cylinder at torn extremity.
G. Non-medullated fiber.

nuclei beneath the neurilemma. The neurilemma is wanting in some situations. The medullated nerve-fibers also lose the medullary sheath when they are about to enter upon their peripheral distribution.

THE NERVE-TRUNKS

The structure of nerve-trunks is most typical in the large nerves composed of medullated nerve-fibers. In transverse sections, medullated nerve-fibers appear like small round cells, in which the axis-cylinders resemble nuclei. The nerve-fibers are collected into bundles called *funiculi*.

The connective tissue, which serves to unite the elements of a nerve-trunk, does not differ materially from the sustentacular tissue of other organs. Different terms are applied, according to its use and location, as follows:

Epineurium.—*Forming the sheath of the entire nerve-trunk.*

Perineurium.—*Surrounding the funiculi composing the nerve-trunk.*

Endoneurium.—*Surrounding and uniting the nerve-fibers of the funiculi.*

Neurilemma.—*Surrounding the individual nerve-fibers of a bundle.*

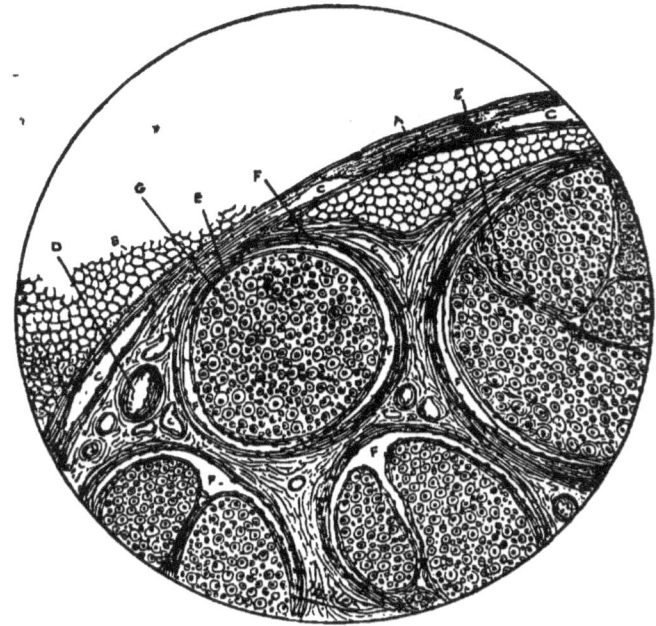

Fig. 138. Transverse Section of the Anterior Crural Nerve (× 250).

A. The epineurium.
B. Adipose tissue in the loose areolar tissue of the sheath.
C. Lymph-spaces of the epineurium.
D. Large blood-vessels of epineurial sheath.
E. Perineurium surrounding funiculi.
F. Lymph-spaces of last.
G. Medullated nerve-fibers in T. S. supported by connective tissue — endoneurium.

The formula E. P. E. N., composed of the initials of the names of the investments from without inward, will aid the memory.

The epineurium serves to protect the nerve in its course, and to support the nutrient blood-vessels and the channels of lymphatics. The fibers run both longitudinally and transversely. The perineurium, arranged in dense bands, forms distinct sheaths for the funiculi, the fibers running, for the most part, circularly. The endoneurium not infrequently divides the nerve-bundles into smaller or primitive bundles. It supports the blood-capillaries, contains small lymph-spaces, and its nuclei are frequently large and prominent. The larger nerve-trunks have their own nerves distributed to the epineurium,—*nervi nervorum*.

The final distribution of the elements of a nerve-trunk is effected by subdivision, first, of the large, and afterward of the primitive bundles or funiculi. The perineurial sheaths are prolonged, surrounding the dividing bundles, even to their final distribution, where, around terminal and single medullated fibers, the sheath remains as a layer of exceedingly delicate flattened cells. The necessity for the endoneurium ceases with the ultimate subdivision of the funiculus.

The medullated nerve-fibers, when they reach the point of their final distribution, branch at a node of Ranvier, and also lose the medullary sheath.

PRACTICAL DEMONSTRATION

Medullated nerve-fibers can be studied best in preparations that have been teased on a slide. It requires much care and patience to separate the single nerve-fibers of a bundle. The student should use nerve-tissue that was fixed with osmic acid while fresh (see page 23). The medullary sheaths are black. Also, tease nerve-fibers that were hardened in Müller's fluid and afterwards washed and transferred to alcohol. Stain the fibers on the slide with Van Gieson's picric acid and acid fuchsin mixture (page 32), omitting the hæmatoxylin. Dehydrate with a few drops of alcohol in succession, which may be removed with blotting-paper; clear with a drop of oil of cloves; mount in balsam. The axis-cylinders and nuclei are red; the medullary sheaths are yellow; the nodes of Ranvier are distinctly visible.

A nerve that has been hardened in Müller's fluid should be imbedded in celloidin, and transverse sections cut; stain with hæmatoxylin and Van Gieson's picric acid and acid fuchsin.

NERVE-CELLS

The cells of the nerve-centers are usually called *ganglion-cells*. They differ greatly in size, some of the largest measuring $100\,\mu$ or more in diameter. The nucleus is round, conspicuous, and has a nucleolus. The protoplasm sometimes contains pigment. The cells are surrounded by minute lymph-spaces. Ganglion-cells exhibit various shapes, some being spherical, others pyriform or stellate. The differences in outline are due, in part, to the processes, of which there may be one or many. According to the number of processes, the cells are sometimes named unipolar, bipolar, or multipolar. Every ganglion-cell has a relatively straight process, which is the *axis-cylinder process*. The other processes, when they are present, divide and subdivide rapidly, to form fine networks or arborizations. They are called *protoplasmic processes*,

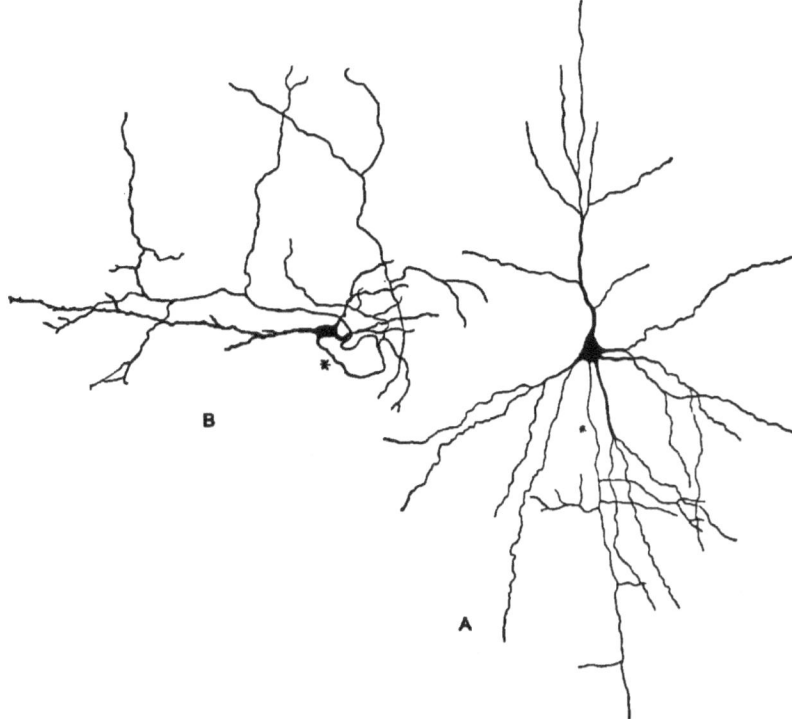

Fig. 139. Types of Ganglion-cells as shown by Golgi's Method. (Baker.)
A. Type I. * Axis-cylinder process with collaterals.
B. Type II. * Axis-cylinder process.

or *dendrites*. The development of protoplasmic processes is seen in an extreme degree on the ganglion-cells of Purkinje in the cerebellum (Fig. 158). It is probable that the processes of one cell do not unite with those of any other, and that whatever physiological relations the cell has with other cells are established through proximity, but not by continuity of substance. The researches made according to Golgi's method of impregnating tissues with silver have shown that ganglion-cells are of two principal types:

NERVE-CELLS 221

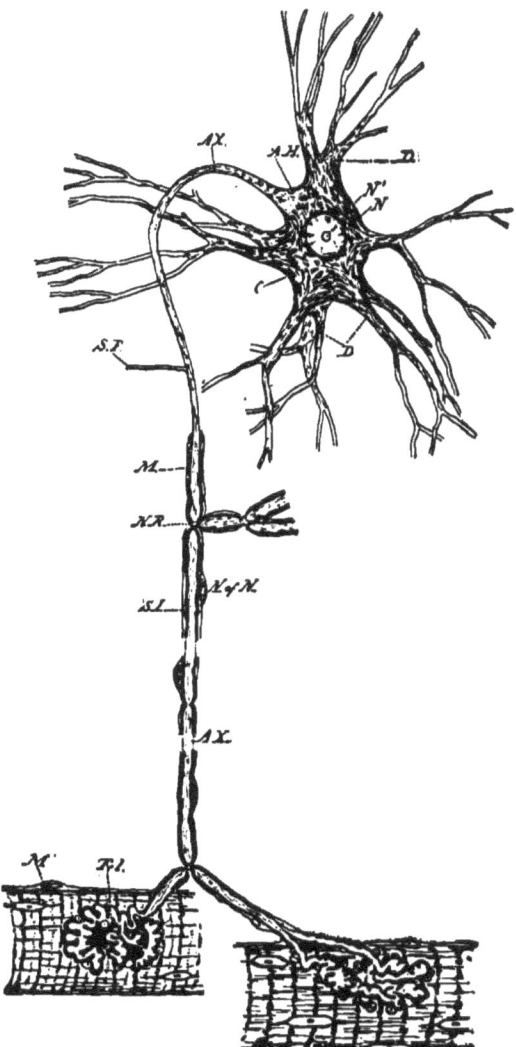

FIG. 140. DIAGRAM OF THE LOWER MOTOR NEURONE. (BARKER.)

D. A motor-cell from the anterior horn of the spinal cord, with its protoplasmic processes.
Ax. Axis-cylinder process.
S. F. Collaterals.
M. Medullary sheath.
N. R. Node of Ranvier.
N. of N. Nucleus of neurilemma.
Tel. The ending, in striped muscle-fiber, M'.
N. The nucleus, and N' nucleolus of the ganglion-cell.

Type I. The cell has, besides the protoplasmic processes, an axis-cylinder process which becomes continuous with the axis-cylinder of a nerve-fiber usually having a considerable length. Minute offshoots are given from the axis-cylinder process, which are called *collaterals* (Fig. 139, A).

Type II. In this case the axis-cylinder process divides repeatedly and soon after leaving the ganglion-cell, forming a network within the nerve-center (Fig. 139, B).

A ganglion-cell with its axis-cylinder process is called by many writers a *neurone*. In the case of some of the medullated nerve-fibers proceeding from the cerebro-spinal centers, the terminal distribution of the axis-cylinder may be as much as a meter distant from the ganglion-cell (Fig. 140).

NEUROGLIA

The brain and spinal cord have a supporting framework of ordinary connective tissue which enters at the surface from the pia mater. They possess, besides, a special form of supporting tis-

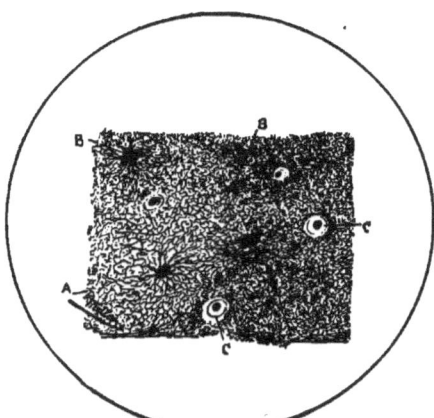

Fig. 141. Neuroglia from beneath the Pia Mater of the Spinal Cord.
(× 400.)

A. Network of neuroglia-fibrils.
B. Spider (Deiter's) cells.
C. Nerve-fibers in T. S.

sue called *neuroglia*. It consists of a fine reticulum produced by the interlacing of large numbers of delicate processes, which arise from neuroglia cells. These cells have a stellate outline, and, with their numerous processes, suggest a spider and its legs (Figs. 139 and 155). Some forms are described as "moss-like." Neuroglia-cells are well shown in silver-stained specimens. The neuroglia of the spinal cord is intimately related with the epithelium lining the central canal, from which it originated. Neuroglia, unlike the other supporting tissues, is derived from the *ectoderm*.

THE PERIPHERAL NERVE-ENDINGS

The terminal branches of the nerves are so complex in structure and so difficult of demonstration that only a few of the most important can be considered in this work.

Sensory nerves sometimes present *free endings*, as in the stratified epithelium of the epidermis and cornea. The medullated nerve-fiber loses its medullary sheath at a node of Ranvier, and divides repeatedly, making a plexus of minute fibrillæ in the connective tissue, or just below the epithelium, or between the epithelial cells.

Special sensory endings are more complex in structure. Examples are found in the *tactile corpuscles*, which occur in the papillæ of the skin. They are oval bodies having numerous nuclei.

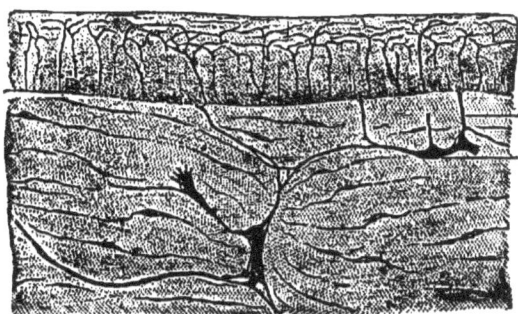

Fig. 142. Termination of Nerve-Filaments in the Epithelium of the Cornea. (Ranvier.)

One or more medullated nerve-fibers enter at the base, losing their medullary sheaths. The axis-cylinders break up into fibrils, which here and there present expansions.

The *Pacinian* bodies or *corpuscles of Vater* are easily found in the mesentery of the cat, where they are numerous and visible to the unaided eye. They are oval in form and nearly transparent. The most prominent part is the capsule, which consists of lamellæ of connective tissue, which are concentric, and resemble the layers

224 STUDENTS HISTOLOGY

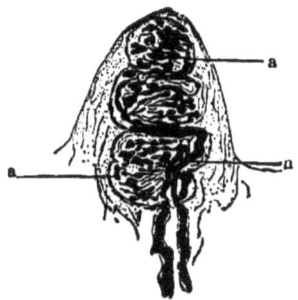

Fig. 143. Tactile Corpuscle from the Skin of the Palmar Surface of the Index Finger of Man. (Ranvier.)
a. Terminal nerve-fibrils.
n. Afferent nerve.

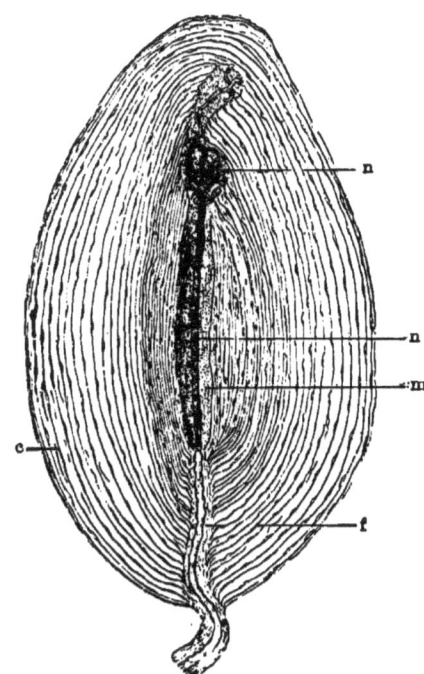

Fig. 144. Pacinian Corpuscle from the Mesentery of the Cat. (Ranvier.)
c. Capsule.
m. Inner bulb.
f, n. Afferent nerve.

of an onion. A medullated nerve-fiber enters at the bottom, and soon loses its medullary sheath. The free axis-cylinder terminates in a rounded or bulb-like extremity, surrounded by a homogenous substance, the inner bulb.

The special form of nerve-endings, called *taste-buds*, have already been described (page 149).

NERVE-ENDINGS IN MUSCLE

The plexuses of non-medullated nerves supplying non-striated muscle have often been referred to (pages 154 and 160). From these plexuses minute fibrillæ extend among the muscle-cells. Their exact mode of termination has not been definitely determined.

FIG. 145. NERVE-ENDING IN STRIATED MUSCLE—AFTER KÜHNE.

The medullated nerves supplying striated muscle form an intramuscular plexus. Bundles of nerve-fibers start from this plexus, one nerve-fiber going to each muscle-fiber. The medullary sheath is lost. The axis-cylinder breaks up into twisted fibrillæ, with bulbous ends. These constitute the *end-plate*, which probably lies below the sarcolemma, and is imbedded in nucleated protoplasm, the *sole-plate*.

SPINAL CORD

The membranes covering the spinal cord will be discussed later; see page 235.

The spinal cord is composed of gray matter (cellular) and white matter (consisting of nerve-fibers), and serves as a medium of communication between the brain and the peripheral nerve-apparatus. The arrangement of its several parts will be best understood by the study of a transverse section, of which Fig. 146 is a diagrammatic representation.

The gray substance occupies the central portions of the struc-

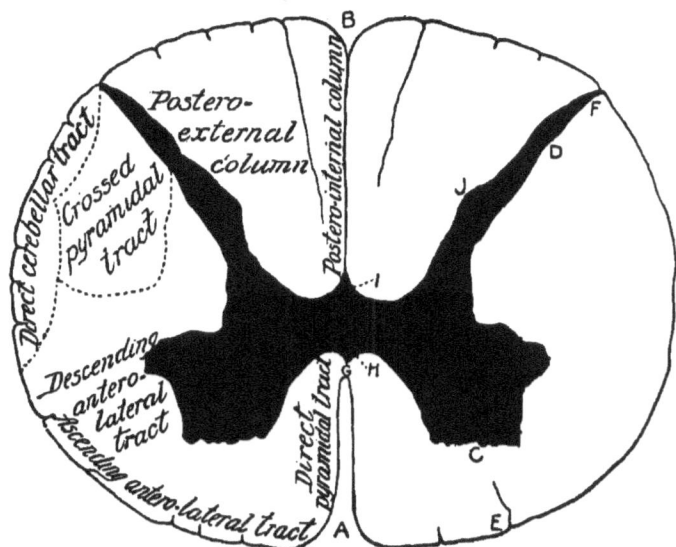

FIG. 146. DIAGRAM. CERVICAL SPINAL CORD IN TRANSVERSE SECTION.

- A. Anterior median fissure.
- B. Posterior median fissure.
- C. Anterior or ventral horn.
- D. Posterior or dorsal horn.
- E. Point of emergence of anterior root of spinal nerve.
- F. Posterior root of spinal nerve.
- G. White commissure.
- H. Anterior gray commissure.
- I. Posterior gray commissure.
- J. Substantia gelatinosa Rolandi.

The tracts which are named on the diagram have no definite boundaries histologically. They are physiological areas.

ture, and consists of two lateral masses and a connecting link, or commissure. Near the central portion of the figure, a small circular opening occurs—the transversely divided *central canal*.

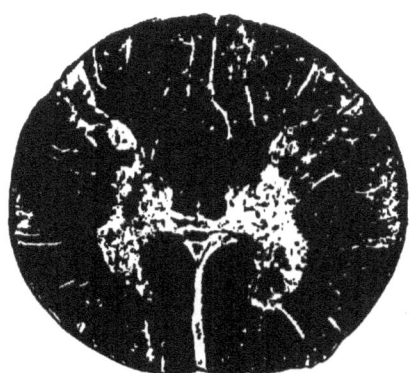

FIG. 147. HUMAN SPINAL CORD FROM THE DORSAL REGION, STAINED BY THE WEIGERT-PAL METHOD. SLIGHTLY MAGNIFIED. PHOTOMICROGRAPH.

This is in communication, in the medulla, with the fourth ventricle, and will serve as a starting point for our study.

The gray matter completely surrounds the central canal, and its outline resembles the capital H, or a pair of crescents with their concavities looking outwards. The anterior (or ventral) horns or cornua are blunt. The posterior (or dorsal) horns are pointed. There are lateral projections from the anterior horns, sometimes called lateral horns. They lie opposite the central canal, and are most marked in the upper thoracic region. The crescents are connected, a portion of the connecting substance passing in front and a portion behind the central canal—*the anterior and posterior gray commissural bands*. The amount of gray matter is greatest in the cervical and lumbar enlargements of the cord (Fig. 148).

The white substance is divided anteriorly by the *anterior median fissure*, which cuts into the cord nearly to, but not quite as far as the anterior gray commissure. A corresponding division appears posteriorly (the *posterior median fissure*), which does not divide the cord posteriorly, but the division is indicated by a band of pia mater, which penetrates entirely to the posterior gray commissure. The two masses of white substance thus indicated are roughly divided into anterior, lateral, and posterior columns by the horns of gray matter. They are united just in front of the anterior gray commissure by white matter—the *white commissure*. The *spinal nerves* take origin from the gray cornua, the *anterior roots* from the anterior, and the *posterior roots* from the posterior cornua. The white substance consists essentially of medullated nerve-fibers which, with the exception of the anterior spinal nerve-roots and the commissural fibers, pass mainly in a longitudinal direction.

The anterior, lateral, and posterior columns into which the white matter is primarily divided, may be divided secondarily

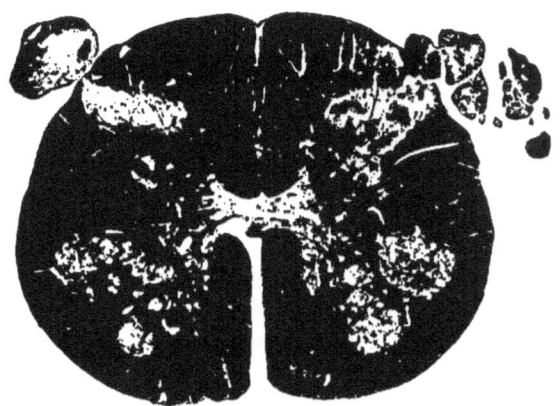

Fig. 148. Human Spinal Cord from the Lumbar Region. Stained by the Weigert-Pal Method. Slightly Magnified. Photomicrograph.

into certain other columns or tracts which are indicated in Fig. 146. These tracts are often called by the name of the discoverer. The principal ones are as follows:

The direct pyramidal tract: Türck.

The ascending antero-lateral tract: Gowers.

The descending antero-lateral tract.

The crossed pyramidal tract.

The direct cerebellar tract.

The postero-external tract—funiculus cuneatus: Burdach.

The postero-internal tract—funiculus gracilis: Goll.

The demonstration of the different tracts we owe partly to pathology, since in certain diseases definite tracts may be involved throughout the length of the cord, while the others may be exempt. The alterations which ensue in the diseased parts make it easy to trace such tracts.

Embryological study has shown also that the medullary sheath appears at different periods of development in the different tracts of nerve-fibers, although the time is constant for each particular tract. The presence of the sheath in certain tracts, and its absence in others, makes it possible to outline the tracts in a series of embryonic cords.

Physiological experiments demonstrate the functions belonging to certain tracts.

PRACTICAL DEMONSTRATION

Nerve-tissue should, under all circumstances, be hardened in Müller's fluid. The cord should be obtained as nearly fresh and uninjured as possible, cut transversely with a sharp razor into pieces a centimeter long, and placed immediately in the fluid—in the proportion of a liter of the mixture to 100 grams of tissue. The solution should be thrown away after twenty-four hours, and a fresh supply provided. It should be again changed after three days, and again after another week. After four weeks the bichromate should be poured off, and the tissue rinsed once with water, after which the hardening is to be completed with alcohol in the ordinary manner.

After hardening, pieces from the different regions should be cut, and this is best effected after imbedding. Transverse sections are the most instructive, although the student should afterward study longitudinal sections. The sections may be stained with hæmatoxylin and eosin. Sections should also be stained by the Weigert-Pal method. For this purpose the hardening must be done with great care, and must be continued longer. The pieces of tissue are placed in alcohol direct from Müller's fluid without washing. The details of the method are given on page 30. The Golgi method presents too many difficulties to be attempted by any but advanced students.

HUMAN SPINAL CORD—CERVICAL REGION—TRANSVERSE SECTION (Fig. 149)

OBSERVE :

(L.)

1. General **arrangement of gray and white substance**, with the latter surrounding the former, which resembles in outline the letter H.

2. **Subdivisions of white substance.** (*a*) **Anterior median fissure.** (Note its passage inward and its cessation before reaching the gray substance.) (*b*) **Posterior median fissure.** (Note its shallowness as a true fissure, and the extension of the connective tissue from the bottom inward, until the gray substance is met. Compare the two median fissures.) (*c*) The emergence of the **anterior nerve-roots.** (This provides the external or lateral

boundary of the anterior white columns, the internal boundaries being provided by the anterior median fissure.) (*d*) The **lateral columns**. (These contain the fibers of the crossed pyramidal tract, and include the white substance between the anterior nerve-roots and the posterior gray cornua. Each lateral column contains nerve-fibers which pass to the cerebellum—direct cerebellar tract; observe that these tracts have no internal histological boundary. Note the numerous prolongations of the pia mater inward in the lateral columns and blood-vessels in them. (*e*) The **postero-in-**

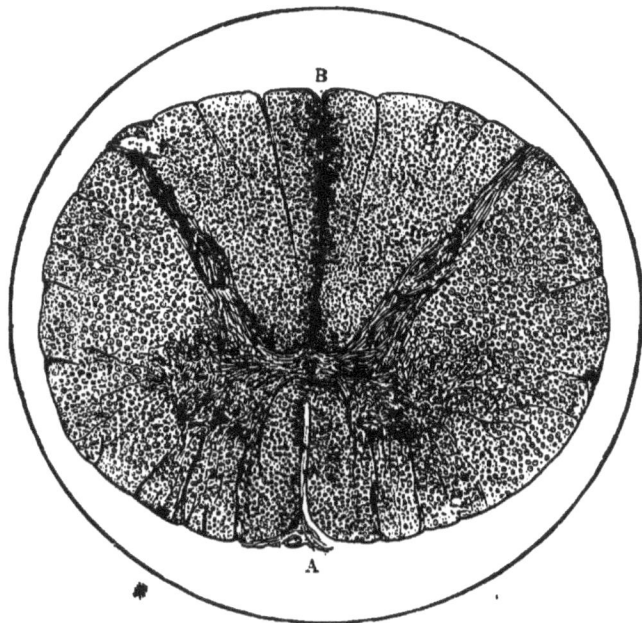

Fig. 149. Transverse Section of the Spinal Cord. Middle Cervical Region (× 60).

A. Anterior. B. Posterior.

This section was made from the cord of a man who died at the age of 75 years, from senile dementia. The gray substance appeared normal, but of somewhat diminished area.

ternal or **column of Goll**—*funiculus gracilis*. (These columns occur on either side of the posterior median fissure, and are bounded laterally by a prolongation from the pia mater.) (*f*) The **postero-external column**—*funiculus cuneatus*. (Bounded internally by the postero-internal column, and externally by the posterior gray cornu.) (*g*) The **white commissure**. (Note the absence of a white commissure posteriorly, the posterior median septum reaching the gray substance.)

3. **Subdivisions of the gray substance.** (*a*) The **central canal.** (Its size and shape vary a good deal at different levels.) (*b*) The **gray commissures,** anterior and posterior. (*c*) The **gray columns.** (*d*) The **anterior gray cornua,** broad and not reaching the periphery of the cord section. (*e*) The **posterior cornua,** narrow and passing completely out, posteriorly, to form the posterior root of a spinal nerve.
(H.)

4. The **white substance.** (Select a field, e. g., in the anterior

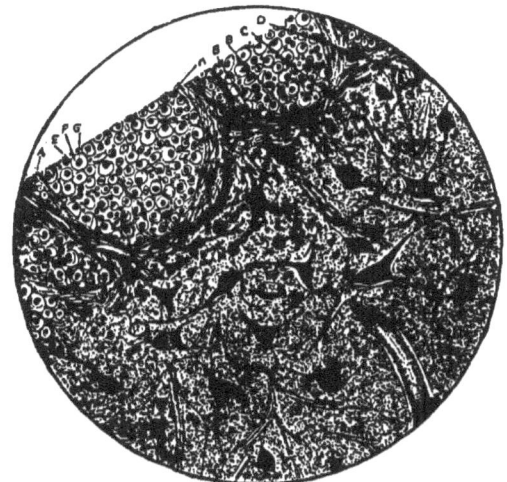

FIG. 150. SAME SPECIMEN AS SHOWN IN FIG. 149. MORE HIGHLY MAGNIFIED. REGION OF ANTERIOR CORNU (× 350).

A. Medullated filaments passing out from the gray substance to form the anterior root of a spinal nerve.
B. Ganglion-cells.
C. Neuroglia-nuclei.
D. Blood-vessels.
E. One of the transversely divided medullated fibers of the white substance, anterior to the anterior gray cornu. The line leads to the neurilemma.
F. White substance of Schwann—of E.
G. The axis-cylinder of E.

median column, and observe the transversely divided nerves.) (*a*) The nerves are not collected into funiculi, but **each fiber pursues an independent course.** (*b*) The **axis-cylinders,** which suggest cell-nuclei. (Note the great variation in size.) (*c*) Most of the axis-cylinders surrounded by more or less concentric rings of translucent, unstained **white substance of Schwann.** (These are medullated fibers. In the spinal cord a neurilemma is wanting for these fibers. With the Weigert-Pal method the medullary sheaths are stained black and their course is easily traced.) (*d*) The small deeply hæmatoxylin-stained **cells of the neuroglia.** (*e*) The **neuroglia-substance,** finely granular or fibrillated, between the nerve-fibers. (*f*) The **spider cells** (Deiter's) of the neuroglia. (These are not numerous, but easily found near the periphery.) (*g*) The **longitudinal nerve-fibers** passing from the anterior gray cornu to form the anterior root of a spinal nerve. (*h*) The different **size of the nerve-fibers in different areas** of the section. Note the small fibers of the postero-internal column. With certain exceptions, the larger fibers belong to motor tracts and the smaller fibers to sensory tracts. (*i*) The **blood-vessels.** (These vessels are largely confined to the fibrous septa, which pass in from the pia.)

5. The **gray substance.** (*a*) The **central canal.** (The canal is lined with columnar ciliated cells in a single layer. The cilia are rarely demonstrable in the human cord, except in children. The central canal in adults is often partly occluded. Observe the clear, homogeneous ground-substance,— **substantia gelatinosa centralis.** Compare it with a similar mass covering the posterior horn,— the **substantia gelatinosa Rolandi.**) (*b*) The **ground-substance.** (This consists, first, of exceedingly minute fibers, formed by the repeated subdivision of the axis-cylinders—the **primitive fibrillae;** second, of the delicate **neuroglia-fibers.** It is usually difficult in a section to differentiate between the two. The details can only be made out in preparations stained with silver by Golgi's method.) (*c*) **Large ganglion-cells.** (In the anterior horn. The straight, unbranching axis-cylinder process can frequently be distinguished. Note the large, shining nucleus and the deeply stained nucleolus. These cells are frequently deeply pigmented. *They may be divided into two or three separate groups.*) (*d*) **Small ganglion-cells.** (Best seen in the posterior horn. In the dorsal cord a collection of medium sized cells appears at the point where the gray commissure joins the posterior horn,— the column of Lockhart Clarke.) (*e*) The lateral horn also contains small ganglion-cells. Occasional outlying ganglion-cells appear in the white matter of the antero-lateral and posterior columns. (*f*) **Pericellular lymph-spaces.** (Observed as a somewhat clear space around the ganglion-cells.) (*g*) **Blood-vessels.** (These are much more numerous here

than in the white portion; and arteries of considerable size may frequently be found.) (*h*) **Peri-vascular lymphatics.** (Find an artery in transverse section, and observe the clear space around it, which may be mistaken for the result of contraction of the tissue in hardening. Careful study will reveal minute branches of cells, passing between the adventitia of the blood-vessel and the wall of the lymph-space.)

The course of the axis-cylinders in the cord, and their relations with the ganglion-cells of the gray matter, are extremely intricate. The Golgi method, in the hands of Ramon y Cajal and others, has been the means of clearing up many of the obscure parts of this subject. The limits of the present work will only permit of

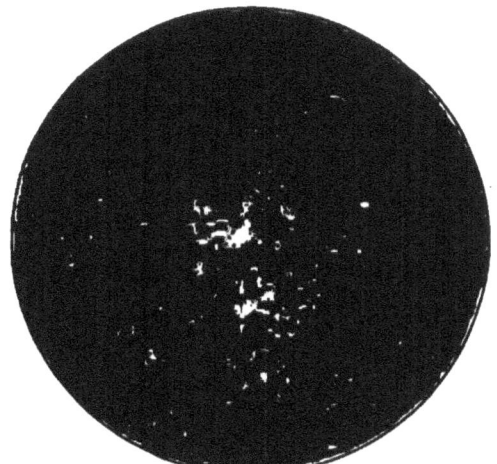

FIG. 151. LARGE GANGLION-CELLS OF THE ANTERIOR HORN OF THE SPINAL CORD. LOW POWER. PHOTOMICROGRAPH.

allusion to a few of the most important facts. In this connection, it is well to recall the description and diagram of a neurone (page 222). The central nervous system is supposed to consist of many neurones. According to this theory, several neurones may operate together, one superimposed on another. It is to be remembered that the processes of ganglion-cells probably do not anastomose with those of other ganglion-cells.

The large ganglion-cells of the anterior horns of gray matter give off axis-cylinder processes to the anterior motor nerve-roots. Some ganglion-cells (column cells) have axis-cylinder processes which run vertically in the white matter of the anterior and lateral

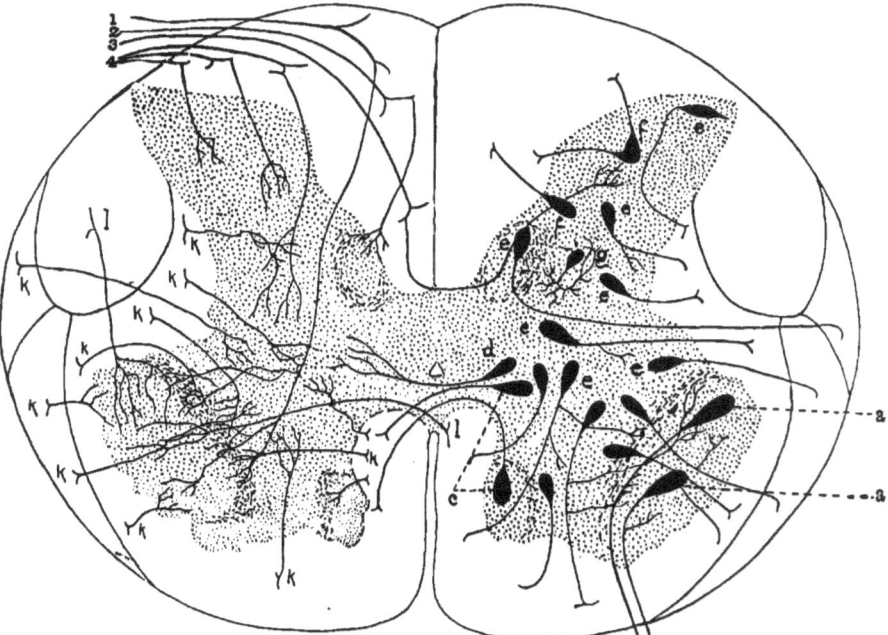

Fig. 152. Diagram of the Relations of the Cells and Fibers of the Spinal Cord. (Baker, after Lenhossek.)

The right side shows the cells of different classes found in the cord, and their processes. The left side gives the processes of cells whose bodies are either beyond the cord or at other levels, with the distribution of their collaterals.

    a, a. Motor cells of the anterior horn.
    c. Commissural cells.
    d. Golgi commissural cell.
    e, e. Columnar cells of antero-lateral column.
    f, f. Columnar cells of posterior column.
    g. Golgi cell of posterior horn.
        1. Fibers of posterior root forming the antero-posterior reflex tract.
        2. Fibers passing to the column of Clarke.
        3. Commissural fibers of posterior root.
        4. Fibers that enter the posterior horn.
    k, k. Collaterals of antero-posterior column.
    l, l. Collaterals from the pyramidal tracts.

columns. Commissural cells give off axis-cylinder processes to the opposite side of the cord, by way of the anterior gray commissure. The axis-cylinders of some ganglion-cells terminate in the gray matter itself. The fibers of the posterior nerve-roots, arising from the cells of the spinal ganglia, divide into ascending and descending branches, which enter the posterior columns.

The difficulties in the way of tracing the paths of the fibers in the cord are enhanced by the numerous collaterals arising from many axis-cylinders.

THE BRAIN AND ITS MEMBRANES

The brain and spinal cord are surrounded by three connective tissue layers—the *dura mater*, the *arachnoid*, and the *pia mater*.

The dura mater is the most external and the thickest of the three membranes, and constitutes the periosteal lining of the cranial cavity. It consists largely of elastic tissue, the laminæ and bloodvessels of which are supported by connective tissue. The outer surface is in more or less intimate connection with the bone; the inner surface is covered with a single layer of thin endothelial cells. Beneath is a space—the *subdural*—containing lymph.

The arachnoid, exceedingly thin, presents an outer glistening surface, covered with a layer of endothelial cells. It is devoid of blood-vessels and nerves. It is separated from the dura by the subdural space, from the under (inner) side of which short, fibrous trabeculæ are projected to the pia. Other trabeculæ attach it loosely to the dura mater. Villous projections from the arachnoid enter the subdural space. Near the longitudinal fissure they are large and encroach upon the dura mater, forming the *Pacchionian bodies*. The *subarachnoidal space* is thus seen to consist of numerous communicating chambers, and these spaces are everywhere lined with flat cells, and contain lymph, as does the subdural space.

The pia mater consists of fibrillated connective tissue, usually in intimate connection with the arachnoid externally, by means of the trabeculæ of the latter. The pia mater is exceedingly vascular, and everywhere covers the brain and cord; and, unlike the arachnoid, penetrates the sulci of the former and the fissures of the latter, becoming continuous with the connective tissue. The outer surface of the pia mater is also covered with flat endothelial cells.

The subdural and subarachnoidal spaces are lymph-cavities, and, while not in direct connection one with the other, belong to the general lymphatic system, and are in eventual connection.

The arrangement of gray and white nerve-substance in the brain is precisely the reverse of that of the cord. The gray matter forms an external covering or layer of varying thickness, while the white matter occupies the more central regions. Collections of gray matter—the basal ganglia—are also situated in the deeper parts of the brain-substance, the study of which does not come within the limits of this work.

The brain-substance does not differ essentially from the cord, except in the arrangement of its parts. The nerve-fibers are

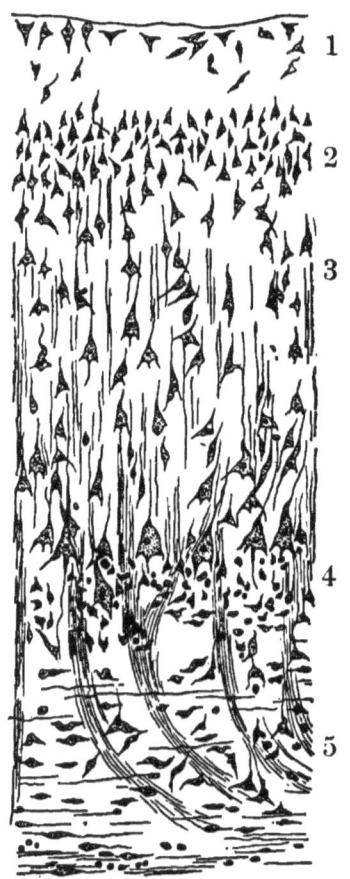

Fig. 153. Layers of the Human Cerebral Cortex. (Meynert.)
The numbers refer to the layers given in the text, page 238.

mostly medullated, but have no neurilemma. The gray substance is arranged in five layers, which are in some instances quite sharply defined, and oftener demonstrable only with considerable difficulty. Transversely running bands of medullated nerve-fibers (the *stripes of Baillarger*) have sometimes been counted as additional layers.

PRACTICAL DEMONSTRATION

The tissue is to be prepared in the manner usual with nerve-substance —hardened with Müller, followed by alcohol. Thin sections, stained deeply with hæmatoxylin and eosin, may be mounted in balsam. Sections stained by Golgi's method should be studied if possible.

SECTION OF HUMAN CEREBRUM—CUT PERPENDICULARLY TO THE SURFACE (Fig. 154)

OBSERVE:

(L.)

The membranes. (In the drawing only the arachnoid and pia are shown.) (*a*) The fine fibrillar mesh of the arachnoid. (*b*) The nuclei of the flattened cell-covering. (*c*) The large

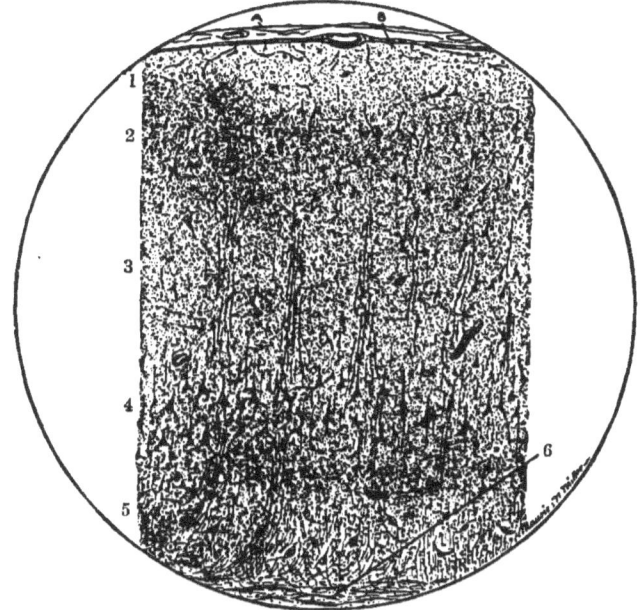

FIG. 154. VERTICAL SECTION OF CEREBRAL CORTEX. SUPERIOR FRONTAL CONVOLUTION (× 250).

A. Arachnoid.
B. Pia mater.
1, 2, 3, 4, 5. First, second, third, fourth, and fifth layers of gray matter.
6. White matter.

blood-vessels. (*d*) The pia. (*e*) Its continuity with the connective tissue of the cerebrum.

1. **The outer layer**—the first—of the gray substance. (This layer is poorly defined, but can usually be made out. It consists of primitive nerve-fibrillæ, neuroglia-fibrils, and scattered ganglion-cells. A few medullated nerve-fibers run horizontally—**tangential fibers.**)

2. **The second layer.** (This layer presents about the same thickness as the preceding, and will be recognized by the abundance of small triangular nerve-cells. From Golgi preparations it appears that numerous protoplasmic processes pass peripherally, while the axis-cylinders arise from the bases of the **small pyramidal ganglion-cells.**)

3. **The third layer.** (This layer—the thickest of all the gray laminæ—is called the **formation of the cornu Ammonis** [Meynert]. The **large pyramidal cells** have numerous protoplasmic processes arising from their sides and prominent ones from the apices, while the axis-cylinders are given off from the blunt bases to enter the white matter [Fig. 155]. Medullated fibers, in more or less distinct bundles, pass between the column-like ganglion-cells.)

4. **The fourth layer.** (The large cells of the third layer are seen to stop, as we pass inward, and give place to small, irregular nerve-cells, called the granular formation. Between the cells of this layer bundles of nerve-fibers are seen, as they radiate toward the cerebral surface.)

5. **The fifth layer.** The line of demarcation between this and the fourth layer is feebly shown; but, on close attention, it will be observed that the small cells of the fourth layer rather abruptly give place to spindle-shaped ones, sometimes parallel with the surface. The nerve-bundles are here more plainly indicated.

6. **The white matter.** (The ganglion-cells cease here, and the field is occupied with medullated fibers and neuroglia, the spherical nuclei of the latter becoming prominent from the deep hæmatoxylin staining.)

7. **The nutrient blood-vessels.** (The capillaries projected from the pia are especially to be noticed, often of the diameter of a single blood-corpuscle, and appearing as branching lines, composed of these elements—indeed difficult of demonstration when empty. Note the light, perivascular lymph-spaces, well seen around the larger arteries in transverse section.)

Certain concentrically striated bodies, **corpora amylacea**, are often found in the vicinity of the ventricles and along the olfactory tract, as well as in other localities. After treatment with iodine solution and sulphuric acid they take a violet color, therefore resembling starch in their reactions—hence their name.

The arrangement of the layers of the gray matter is subject to considerable modification in different parts of the cerebrum. As in the spinal cord, the difficulties of tracing the course of the nerve-

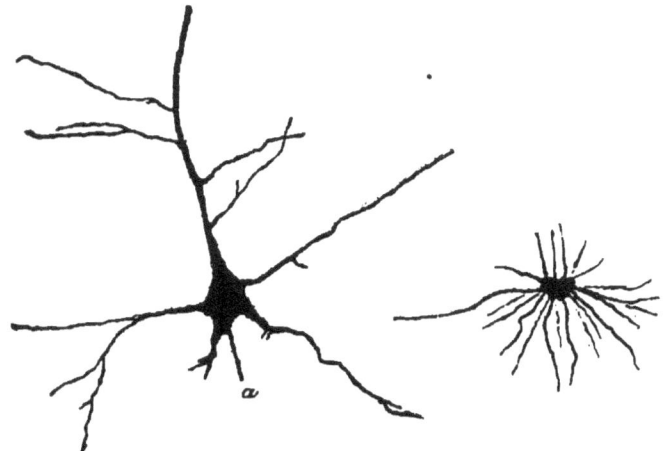

Fig. 155. Pyramidal Ganglion-Cell and Neuroglia-Cell from Human Cerebral Cortex Prepared by the Golgi Method.

a. Axis-cylinder process of the ganglion-cell.

fibers are increased by the numerous collaterals given off from the axis-cylinder processes of the ganglion-cells.

The paths followed by the fibers of the white matter are very complicated. The tracts have been mapped out largely by preparations made by Weigert's method and its modifications. In general these fibers may be divided into three groups:

a. **Projection fibers**, which constitute the corona radiata, passing from the peduncles of the cerebrum and the basal ganglia to the cortical gray matter.

b. **Association fibers**, which bring parts of the same hemisphere into relation with one another.

c. **Commissural fibers**, which connect the opposite hemispheres through the corpus callosum and the anterior commissure.

VERTICAL SECTION OF HUMAN CEREBELLUM

Practical Demonstration

Cut large sections of cerebellum, hardened with Müller's fluid and imbedded as usual; cut so as to show the ramifications of the *arbor vitæ*; stain with hæmatoxylin and eosin. Also stain the cerebellum of a young dog or kitten, prepared according to Golgi's method (page 33), being careful to cut the sections at a right angle to the long axis of the convolutions.

OBSERVE: (Figs. 156 and 157)
(L.)

1. The **arrangement** in the form of leaflets.

2. The **extension of the gray laminae** within even the minutest folds of the leaves, so as to completely envelop the central white nerve-substance. (The staining has been so selected by the

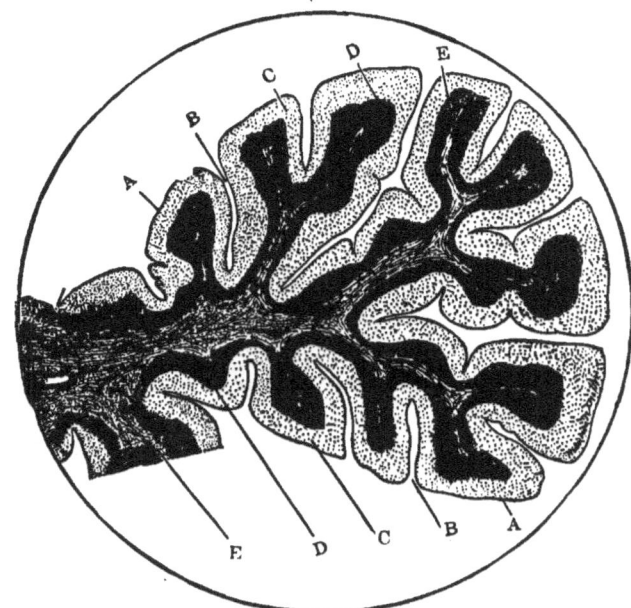

FIG. 156. LONGITUDINAL SECTION OF ONE OF THE FOLIA OF THE CEREBELLUM (×60).

    A, A. Line of pia mater.
    B, B. Sulci.
    C, C. Outer layer of gray matter.
    D, D. Inner layer of gray matter, including Purkinje's cells.
    E, E. White nerve-substance.

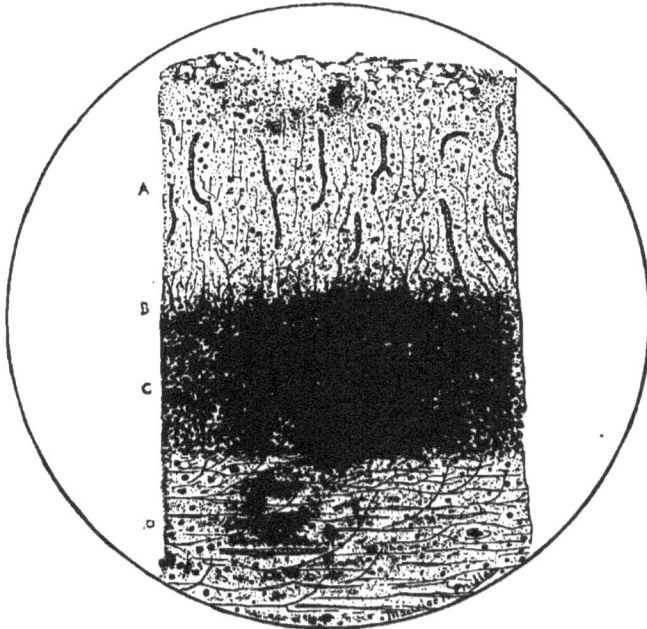

Fig. 157. Vertical Section, Cortex of Cerebellum. Portion of Section shown in Fig. 156, more highly magnified ($\times$ 250).

- A. Outer layer of gray matter.
- B. Layer of Purkinje's cells.
- C. Inner gray layer.
- D. White nerve-substance.

tissue as to divide the outer gray matter into two prominent layers. The explanation of this will follow increased amplification.)

3. The **central white matter**. (The fibrillar character can be made out, and the general plan will be found to consist, as in the cerebrum, of central nerve-fibers radiating toward the cells of the cortical gray substance, the **arbor vitae**.)

(H.)

4. The **outer gray layer, or molecular layer**. (This is the thickest of the three layers. The prominent elements to be observed are: the **scattering neuroglia- and ganglion-cells, nerve-fibrils**, and **blood-vessels**, which pass in from the pial investment.) The ganglion-cells of the molecular layer are of two sorts, small and large. The large cells are remarkable for their axis-cylinder processes, which at intervals give off branches whose

fine subdivisions form a basket-like network about the bodies of the Purkinje cells of the underlying layer. The arborizations of these Purkinje cells spread out in the molecular layer.

5. The **middle layer** is a thin stratum directly beneath the outer layer. The section becomes deeply stained, from the presence of numerous small cells, among and partly concealed by which are the very large **ganglion-cells of Purkinje**. These are flask-shaped, and are arranged in a single plane, with their long axes placed vertically. A thread-like prolongation may be seen

FIG. 158. CELL OF PURKINJE. HUMAN CEREBELLUM. GOLGI METHOD.

penetrating the layer beneath, providing the cell has been centrally sectioned. This is the axis-cylinder process, which gives off certain collaterals and becomes the axis-cylinder of a medullated nerve-fiber. Two thick protoplasmic processes project from the outer end of the cell, which ramify in the molecular layer. Their arborizations are luxuriant, and the branchings of the various cells have been likened to a forest. Considering the large size of the cells of Purkinje, their axis-cylinder processes, the richness of the branches of their protoplasmic processes, and the curious basket-work of nerve-fibrils around the cell-bodies, they are among the most striking of the objects that have been revealed by the Golgi method. (Fig. 158.)

The branches of these cells spread out chiefly on planes at right angles to the long axis of the convolution. Sections should therefore be made across the convolutions.

6. The **granular layer**. (This is the layer seen so distinctly with the low-power. It consists of innumerable small bodies, deeply stained with hæmatoxylin, usually spherical, which are ganglion- and neuroglia-cells. The ganglion-cells belong to the second type. The majority of these cells are small. A few are large. The small cells have protoplasmic processes which ramify in their own granular layer, and axis-cylinders which terminate in the outer molecular layer. On the other hand the large ganglion-cells of the granular layer have protoplasmic processes which extend into the molecular layer, while the axis-cylinder processes ramify in the granular layer. Medullated nerve-fibers from the white matter form a dense plexus in the granular layer. Some of these fibers are continued into the molecular layer. Search carefully for the **axis-cylinder processes of the Purkinje cells**, which pierce this layer, and follow them into the white matter below.)

7. **The white substance** consists of medullated nerve-fibers.

INDEX

Abbé condenser, 1, 4.
Abdominal cavity. See Peritoneum.
    salivary gland, 145.
Absorption from intestine, 158.
    of fat, 159.
Acid, acetic, 39.
    alcohol, 29, 39.
    aniline dyes, 31.
    chromic, 24.
    fuchsin, 31, 32.
    hydrochloric, 25, 29, 39.
    nitric, 24, 75, 137.
    osmic, 23.
    picric, 23, 29, 31, 32.
Acidophile granules, 31.
Acini of glands, 140.
Acinous glands, 143.
Acoustic nerve, 216.
Adenoid reticulum, 76, 113.
    tissue, 76, 111. See Lymphoid tissue.
Adipose tissue, 65.
Adventitia of blood-vessels, 103, 104.
Agents, staining, 27.
Agminate glands, 160.
Ailantus pith, 13.
Air-bubbles, 46.
    -sacs, 128.
    -vesicles, 128.
Alcohol, acid, 29, 39.
    fixation and hardening, 21.
    dehydration with, 35, 38.
Alimentary canal, 149.
Alum-hæmatoxylin, 28.
Alveoli of gland, 140.
    of lung, 128.
Amœboid movements, 83, 104.
Amphophile granules, 31.
Aniline colors, dyes, etc., 31.
    dyes, staining with, 31.
    oil, 35.
Anterior commissure, 227.
    median fissure, 227.
Antero-lateral tracts, 228.
Aorta, 104.
Appendages of skin, 95.

Appendix, vermiform, 162.
Aponeuroses, 62.
Arachnoid, 235.
Arbor vitæ, 241.
Areolar tissue, 62.
Arrectores pili, 96.
Arrowroot-starch, 48.
Arteries, 102.
    large, 103, 104.
    lymphatics of, 107.
    small, 103.
Arteriolæ rectæ of kidney, 184.
Arterioles, 102.
Artery, bronchial, 128.
    hepatic, 164.
    pulmonary, 128.
    renal, 183.
    typical, 103.
Articular cartilage, 68.
Artificial gastric juice, 25.
Asphaltum varnish, 36.
Association fibers, 239.
Attraction-spheres, 53.
Auerbach's plexus, 154, 160.
Axis-cylinder, 216.
    processes, 220.
    collaterals, 222.

Bacteria, hardening tissues containing, 22.
    under the microscope, 47.
Baillarger's stripes, 237.
Balsam, Canada, 35.
Basement membranes, 55.
Basic aniline dyes, 31.
Basophile granules, 31, 87.
Beaker cells. See Goblet cells.
Bellini, tubule of, 182.
Bergamot oil, 35.
Bertini, columns of, 179.
Berlin blue, injecting, 34.
Bichromate of potassium, 22.
Bile-capillaries, 166.
    -ducts, injection of, 176, 177.
Bipolar nerve-cells, 220.
Birds, blood of, 82.
Bladder, gall-, 177.

Bladder, urinary, 194.
Blastodermic layers, 54.
Blood, 81.
    colorless corpuscles, 83.
    -corpuscles as a standard of measurement, 8.
    -corpuscles, cover-glass preparations, 84.
    -corpuscles, crenation of, 82, 90.
    -corpuscles, double staining, 84.
    -crystals, 89.
    effect of reagents upon, 90.
    effects of acetic acid, 91.
    effects of osmic acid, 82.
    effects of tannic acid, 91.
    effects of water, 90. [32, 85.
    Ehrlich-Biondi-Heidenhain stain,
    embryonal origin of, 91.
    enumeration of corpuscles, 87.
    fibrin, 90.
    fixing and staining, 84.
    hæmin, 90.
    hæmoglobin, 89.
    hæmosiderin, 89.
    hæmatoidin, 89.
    human, 81.
    leucocytes, 83.
    of birds, 82.
    of camelidæ, 82.
    of fishes, 82.
    of frog, 90.
    of invertebrates, 82.
    of lamprey, 91.
    of reptiles, 82.
    origin of colored corpuscles, 91.
    origin of white corpuscles, 87.
    oxygenation of, 89.
    -plates, 82.
    red corpuscles, 81.
    size of corpuscles, 81.
    tricolor or triacid stain, 32, 85.
    -vessels, 102.
    -vessels, capillary, 103.
    -vessels, development of, 105.
    -vessels, injection of, 34.
    white corpuscles, 83.
Bodies, Malpighian of kidney, 180.
    Malpighian of spleen, 118.
    Pacchionian, 235.
    suprarenal, 213.
    thymus, 121.
    thyroid, 148.
    of Langerhans, 148.
Bone, 70.
    circumferential lamellæ, 72.

Bone, compact, 72.
    cancellous, 72.
    -corpuscles, 72.
    decalcification of, 24, 75.
    development of, 73.
    Haversian canals, 71.
    Haversian lamellæ, 72.
    interstitial lamellæ, 72.
    -lacunæ, 70.
    -marrow, 73.
    osteoblasts, 73.
    osteoclasts, 73.
    perforating fibers of Sharpey, 71.
    periosteum, 73.
    spongy, 72.
    varieties of, 72.
Borax-carmine, 29, 39.
Boundary region of kidney, 187.
Bowman, capsule of, 180.
Bowman's muscle-disks, 80.
Brain, 235. See Cerebrum.
Branched tubular glands, 143.
Bronchial artery, 128.
    tube, 123.
Bronchi, 123.
Bronchus of pig, 125.
Brownian movements, 46.
Brunner's glands, 157.
Bubbles, air-, 46.
Buccal epithelium, 56, 149.
    glands, 143.
    cavity, 149.
Burdach's column, 228.

Calcified cartilage, 74.
Calcification, 73.
Calyces of kidneys, 178.
Canada balsam, 35.
Canal, alimentary, 149.
    central, 227.
    dentinal, 135.
    Haversian, 71.
    of the epididymis, 210.
    portal, 165.
Canaliculi, 70.
Cancellous bone, 72.
Capillaries, blood-, 103.
    bile-, 166.
    lymphatic, 106–110.
Capillary bronchial tubes, 123.
Capsule of Bowman, 180.
    of glands, 140.
    of Glisson, 163.
    of kidney, 178.
    of lymph-nodes, 112.

Capsule of spleen, 117.
 of thymus body, 121.
 suprarenal, 213.
Carbol-turpentine, 35.
 -xylol, 35.
Carbon dioxide freezing, 20.
Cardiac glands, 152.
 muscle-fiber, 80.
Care of miscrocope, 43.
Carmine and picric acid, 29.
 Grenacher's borax-, 29.
 injection, 34.
 staining, 39.
Cartilage, 67.
 articular, 68.
 arytenoids, 123.
 -corpuscles, 67.
 elastic, 68, 123.
 fibro-, 68.
 hyaline, 67.
 -lacunæ, 67.
 of epiglottis, 123.
 of larnyx, 123.
 of Santorini, 123.
 of trachea, 68, 123.
 of Wrisberg, 123.
 perichondrium of, 67.
 -plates in bronchi, 124.
 reticular, 68, 123.
 varieties of, 67.
Cavity, abdominal. See Peritoneum.
 pericardial. See Pericardium.
 thoracic or pleural. See Pleura.
Cedar-wood oil, 4.
Cell, 49.
 -cement, 50.
 distribution, 50.
 division, 51.
 typical, 49.
 -wall, 49.
Celloidin imbedding, 26.
Cells, acidophile, 31.
 amphophile, 31.
 basophile, 31, 87.
 bipolar nerve-, 220.
 blood-, 81.
 border, 153.
 central, 153.
 chief, 153.
 Deiter's, 222.
 endothelial. See Endothelium.
 eosinophile, 31, 86.
 epithelial. See Epithelium.
 ganglion-, 219.
 giant. See Giant cells.

Cells, goblet. See Goblet cells.
 lymphoid. See Lymphoid cells.
 mucous. See Goblet-cells.
 multipolar nerve-, 220.
 nerve-, 219.
 neutrophile, 31, 86.
 neuroglia, 222.
 of Purkinje, 242.
 outlining of by silver nitrate, 33, 60.
 parietal, 153.
 pigment-, 93.
 prickle, 92.
 spider, 222.
 typical, 49.
 unipolar nerve-, 220.
 wandering, 83, 104.
Cement-substance, 62, 104, 108.
 of tooth, 137.
 zinc, 36.
Central canal of spinal cord, 227.
 nervous system, 226-243.
Centrosome, 53.
Cerebellar tracts, direct, 228.
Cerebellum, 240.
 cells of Purkinje, 242.
 granular layer, 243.
 molecular layer, 241.
Cerebral cortex, layers of, 238.
Cerebrum, 236.
 association fibers, 239.
 blood-vessels, 238.
 commissural fibers, 239.
 nerve-fibers of, 239.
 practical demonstration, 237.
 projection fibers, 239.
 tangential fibers, 238.
 pyramidal cells of, 238.
Cervix uteri, 198.
Chalice cells. See Goblet cells.
Chamois leather, 43.
Channels, lymph-. See Lymphatics.
 blood-. See Blood-vessels.
Chloroform, 25.
Chromatin, 28, 51).
Chromosomes, 52.
Chromic acid, fixing and hardening, 24.
Chromic-acetic solution of Flemming, 23.
 -osmic solution of Flemming, 23.
Chyle-receptacle, 106.
Ciliary motion, 59.
Ciliated cells from oyster, 59.
Circulatory system, 102.
 system, lymphatic, 106.
Circumferential lamellæ, 72.

Circumvallate papillæ, 149.
Clark's columns, 232.
Classification of tissues, 53.
Cleaning cover-glasses, 41.
　　　　lenses, 44.
Clearing agents, 35, 38.
Clove-oil, 35.
Coarse adjustment, 1.
Cohnheim's fields, 79.
Coiled tubular glands, 140.
Collaterals of axis cylinders, 222.
Collecting tubule, 182.
Collodion, 26.
Colloid substance, 148.
Colorless blood-corpuscles, 83, 104.
Colors, aniline, 31.
Colostrum-corpuscles, 208.
Columns of the spinal cord. See Spinal cord.
Columnar epithelium, 57.
Columns, cortical, of Bertini, 179.
Commissural fibers, 239.
Commissure, anterior gray, 227.
　　　　posterior gray, 227.
　　　　white, 228.
Compact bone, 72.
Compound acinous glands, 143.
　　　　racemose glands, 143.
Condenser, Abbé, 1, 4.
Coni vasculosi, 210.
Connective tissue, 62.
　　　　tissue, embryonic, 76.
　　　　tissue, juice-canals of, 106.
　　　　tissue, mucoid, 76.
　　　　tissue, special, 76.
Conservation of eyesight, 7.
Constrictions of Ranvier, 216.
Convoluted tubule, 181, 182.
Copper sulphate, 24.
Cord, spinal. See Spinal cord.
　　　　umbilical, 76.
Cords of lymphoid tissue, 113.
Corium of skin, 92.
Corn starch, 48.
Cornea, nerves of, 223.
Cornua of spinal cord, 227.
Corpora amylacea of brain, 239.
Corpus callosum, 239.
　　　　Highmori, 210.
　　　　luteum, 203.
Corpuscles, blood-, 81.
　　　　bone-, 70.
　　　　cartilage-, 67.
　　　　colostrum-, 208.
　　　　connective tissue, 63.

Corpuscles, Hassall's, 122.
　　　　lymph-. See lymphoid cells.
　　　　of Vater, 223.
　　　　pus-, 84.
　　　　Pacinian, 223.
　　　　salivary, 47, 149.
　　　　tactile, 94, 223.
Cortex of cerebellum, 241.
　　　　of cerebrum, 237.
　　　　of ovary, 203.
　　　　of kidney, 178.
　　　　of lymph-node, 112.
　　　　of thymus body, 121.
Cortical columns of kidney, 179.
Corundum hones, 16.
Cotton fibers, 48.
Cover-glasses, 41.
Cover-glass, placing of, 42.
　　　　preparations of blood, 84.
Cox's Golgi method, 34.
Crenation of blood-corpuscles, 82.
Creosote, 35.
Crossed pyramidal tracts, 228.
Crusta petrosa, 135.
Crypts of the tonsil, 149.
　　　　of Lieberkühn, 156.
Crystals of Teichmann, 90.
Cul-de-sac of vagina, 198.
Currents, thermal, 46.
Cuticle of hairs, 95.
Cutis anserina, 96.
Cuticula of tooth, 135.
Cutting sections, 10.
Cylinder, axis-, 217.
Cyclostomi, blood of, 91.

Dammar, 36.
Decalcification, 24.
　　　　of bone, 75.
　　　　of teeth, 137.
Dehydrating, 35, 38.
Deiter's cells, 222.
Delafield's hæmatoxylin, 28.
Demilunes of Heidenhain, 148.
Demonstrations, practical. See Practical demonstrations.
Dendrites, 220.
Dentinal canals, 135.
　　　　fibers, 135.
　　　　sheath, 135.
Dentine, 135.
Derivatives of blastodermic layers, 54.
Derma, 92.
　　　　basement membrane, of, 94.
Development of blood-vessels, 105.

INDEX

Development of blood-corpuscles, 91.
    of bone, 73.
Diapedesis, 104.
Diaphragm, central tendon of, 108.
Diaphragm of the microscope, 1.
Direct cell division, 51.
Direct cerebellar tract, 228.
Direct pyramidal tract, 228.
Digestion, method of, 25.
Disk, intervertebral, 68.
    of Bowman, 80.
Discus proligerus, 205.
Dissociating fluids, 10, 25.
Distal convoluted tubule, 182.
Distribution, cell, 50.
Division of cells, 51.
Double staining, 38.
Duct of glands, 140.
    hepatic, 163, 177.
    of thyroid body, 148.
    bile-, 176, 177.
    thoracic, 106, 107.
Ductless glands, 117, 213.
Dura mater, 235.
Dust on lenses, 43.
Duodenum, 162.

Ear, cartilage of, 68.
Ectoderm, derivatives of, 54.
Egg-tubes, 206.   [32.
Ehrlich-Biondi-Heidenhain tricolor stain,
Elastic cartilage, 68.
    fibers, 65.
    lamina of blood-vessels, 102, 103.
    tissue, 63.
Elder-pith, 13.
Eleidin, 92.
Embedding. See Imbedding.
Embryonic tissue, 76.
Enamel of teeth, 137.
    prisms, 137.
Endocardium, 102.
Endochondral formation of bone, 73.
Endomysium, 78.
Endoneurium, 218.
Endothelial cells. See Endothelium.
Endothelium, 60, 103, 108.
    of blood-vessels, 103.
    staining of, 61.
    stomata of, 62.
End-plates, 225.
Entoderm, derivatives of, 54.
Eosin, 28, 31.
    and hæmatoxylin, 38.
Eosinophile granules, 31, 86.

Epiblast, derivatives of, 54.
Epidermis, 92.
Epididymis, 209.
Epiglottis, 68, 123.
Epineurium, 218.
Epithelial cells. See Epithelium.
Epithelium, definition of, 54.
    buccal, 56.
    ciliated, 58.
    classification of, 55.
    columnar, 57.
    distribution of, 55.
    germinal, 203.
    glandular, 59.
    of bladder, 195.
    of ovary, 203, 206.
    of kidney, 182, 192.
    pavement, 56.
    prickle cells, 92.
    simple squamous, 56.
    stratified, 55.
    striated, 148, 192.
    squamous, 55.
    tessellated, 56.
    transitional, 55, 193, 195.
    uterine, 199.
    vaginal, 200.
    varieties of, 55.
Equator of the cell, 52.
Erectile tissue, 212.
Erlicki's fluid, 24.
Erythroblasts, 91.
Ether freezing, 20.
Eustachian tube, cartilage of, 68.
Extraneous substances, 47.
Eye-lens, 3.
    -piece, 3.

Fallopian tube, 202.
Fat, action of osmic acid on, 23.
    -cells, 65.
    -columns, 94.
    -crystals, 66.
    -globules in milk, 45.
    in the liver, 174.
    staining of, 23.
    -tissue, 65.
Feathers, 48.
Female generative organs, 197.
Fenestrated membrane, 65, 103.
Ferrein, pyramids of, 179.
Fibers, association, 239.
    commissural, 239.
    cotton, 47.
    dentinal, 135.

Fibers, linen, 47.
    nerve-, 216.
    perforating of Sharpey, 71.
    projection, 239.
    silk, 47.
    tangential, 238.
    wool, 47.
Fibrin, 90.
Fibro-cartilage, 68.
Fibrous tissue, elastic, 63.
    white, 62.
Field lens, 3.
    of view, 5.
Filiform papillæ, 149.
Fine adjustment, 1, 6.
Fishes, blood of, 82, 91. [227.
Fissure, anterior median, of spinal cord.
    posterior median, of spinal cord,
Fixation of tissues, 20. [227.
Fixing blood-elements, 84.
    fluids, etc., 20.
Flemming's solutions, 23, 24.
Fluid, artificial gastric, 25.
    dissociating, 10, 25.
    Erlicki's, 24.
    Flemming's, 23.
    Müller's, 22.
    Orth's, 22.
    Stirling's, 25.
    Toison's, 87.
Focal adjustment, 5.
Focusing, 5, 45.
Fœtal blood, 91.
    blood-vessels, 105.
    bone, 73.
    lung, 134.
Follicle of hair, 95.
Follicles, Graafian, 203.
    of Lieberkühn, 156.
    of lymphoid tissue, 113.
    solitary, 160.
Foramen cæcum, 148.
Formaldehyde, 22.
Form of objects, 45.
Free nerve-endings, 223.
Freezing microtome, 19.
Fresh tissue, 20.
Frog, blood of, 90.
    capillaries of, 105.
    mesentery of, 60.
    skin of, 56.
Fuchsin, 31.
    acid, 31, 32.
Fungiform papillæ, 149.
Funiculus cuneatus of spinal cord, 228.

Funiculus gracilis of spinal cord, 228.
Funiculi of nerve-trunks, 217.

Gall-bladder, 177.
    -duct, 177.
Ganglia of heart, 102.
    of urinary bladder, 195.
Ganglion-cells. See Nerve-cells.
Gastric fluid, artificial, 25.
    glands, 152.
    tubules, 152.
Generative organs, female, 197.
    male, 209.
Gentian violet, 31.
Germinal epithelium, 203.
    spot, 205.
    vesicle, 205.
Giant cells, 73, 120.
Glands, 140.
    acinous, 143.
    agminate, 160.
    branched tubular, 143.
    Brunner's, 157.
    buccal, 143.
    capsule of, 140.
    cardiac, 152.
    coiled tubular, 140.
    compound racemose, 143.
    compound tubular, 143.
    ductless. See Ductless glands.
    gastric, 152.
    intestinal, 156, 157.
    Lieberkühn, 156.
    lenticular, 156.
    Littré, 196.
    lymphatic. See Lymphatic nodes.
    mammary, 208.
    mucous. See Mucous glands.
    mucous, of bronchi, 127.
    Naboth's, 199.
    pancreas, 145.
    parenchyma of, 140.
    parotid, 143.
    peptic, 152.
    Peyer's, 160.
    pyloric, 152.
    racemose, 143.
    salivary, 143, 148.
    sebaceous, 98.
    simple tubular, 140.
    sublingual, 148.
    solitary, 160.
    submaxillary, 144.
    sudoriferous, 97.
    sweat-, 97.

INDEX

Glands, tubular, 140, 143.
    thyroid, 148.
Glandular epithelial cells, 59.
Glassy membrane of hair, 95.
Glisson's capsule, 163.
Globus major, 209.
    minor, 209.
Globules, oil-, 46.
Glomerulus of kidney, 184.
Glycerine in mounting, 36, 76.
Glycogen in the liver, 174.
Goblet cells, 58, 126, 127, 157.
Gold-staining, 33.
Golgi's method, 33.
Goll, column of, 228.
Gowers, column of, 228.
Goose-flesh, 96.
Graafian follicles, 203.
Gray commissure, 227.
Gray matter of brain, 236.
    spinal cord, 226.
Grenacher's borax-carmine, 29.
Grübler's dyes, 31.
Gum dammar, 36.
Gun-cotton, 26.

Hæmatoxylin, 28.
    and eosin, 38.
    Delafield's, 28.
    staining process, 36.
    Weigert's, 30.
Hæmin, 90.
Hæmoglobin, 89.
Hæmocytometer, 87.
Hæmatoidin, 89.
Hæmosiderin, 89.
Hair, 95, 99, 101.
    -follicles, structure of, 95.
    permanent mounting of, 99.
Hardening, 20.
Hassall's corpuscles, 122.
Haversian canals, 71.
    lamellæ, 72.
    system, 71.
Heart, 80, 102.
    blood-vessels of, 102.
    endocardium, 102.
    muscular tissue of, 80.
    nerves of, 102.
    pericardium, 102.
    valves of, 102.
Heidenhain's demilunes, 148.
Henle, loop of, 181.
Hepatic duct, 163, 177.
    veins, 163.

Highmori, corpus, 210.
Histology, definition of, 53.
Hilum of kidney, 178.
    of lymph-node, 113.
    of spleen, 117.
Hones, 16.
Horns of spinal cord, 227.
Horny layer of skin, 92.
Hyaline cartilage, 67.
Hypoblast, derivatives of, 54.

Ileum, section of, 161.
Illumination, 4.
Imbedding with ailantus pith, 13.
    with celloidin, 26.
    with paraffin, 13, 25.
Immersion lenses, 4.
Indirect cell division, 51.
Inflammation, 84, 104.
Infundibula of kidney, 178.
    of lung, 129.
Injection, methods of, 34.
Insects, 48.
Intercellular substance, 50.
Interglobular spaces, 137.
Interlobular veins of liver, 167.
    vessels of kidney, 184.
Intervertebral disks, 68.
Interstitial lamellæ, 72.
Intestines, 151, 156.
    Brunner's glands of, 157.
    Lieberkühn's follicles of, 156.
    mucosa of large, 162.
    mucosa of small, 156.
    Peyer's patches, 160.
    practical demonstrations, 160–
    solitary glands of, 160. [162.
    valvulæ conniventes, 151, 162.
    villi of, 156.
Intima of blood-vessels, 103.
Intramembranous ossification, 73.
Intralobular vein, 163.
Involuntary muscle, 76. See Muscle, non-striated.
Iodine solution, 29.
Iris diaphragm, 1.
Irregular tubule, 182.

Japanese paper, 43.
Juice, gastric, 152.
Jelly of Wharton, 76.
Juice-canals, 106.

Karyokinesis, 23, 51.
Keratin, 92.

Kidney, 178.
    blood-vessels of, 183.
    boundary region of, 187.
    Bowman, capsule of, 180.
    calyces of, 178.
    capsule of, 178.
    collecting tubes of, 182.
    columns of Bertini, 179.
    connective tissue of, 189.
    cortex of, 179.
    labyrinth of, 179.
    diagram of, 179.
    epithelium, 182, 192.
    Ferrein's pyramids, 179.
    glomerulus of, 184.
    Henle's loop, 181.
    hilum of, 178.
    infundibula of, 178.
    labyrinths of, 179.
    lobules of, 178.
    Malpighian bodies of, 180.
    Malpighian pyramids, 178.
    medulla, 179.
    medullary rays of, 179.
    nerves of, 185.
    papillæ of, 178.
    pelvis of, 193.
    practical demonstration of, 186.
    tubes of Bellini, 182.
    tubules, diagram of, 181, 193.
    tubules of, 180.
Knives, sharpening, 16.
Krause's membrane, 79.

Labeling of slides, 42.
Labyrinth of kidney, 179.
Lacteals, 159.
Lacunæ of cartilage, 67.
    of bone, 70.
Lamellæ of bone, 72.
Lamina, internal elastic, 103.
Lamprey, blood of, 91.
Large intestine, 162.
Langerhans, bodies of, 148.
Larnyx, 123.
    cartilages of, 68, 123.
Lateral columns, 228.
Layers of cerebrum, 238.
    of epidermis, 92.
Lens, immersion, 4.
    microscope, 2, 3, 4.
Lenticular glands, 156.
Leucocytes, 83, 104.
Lieberkühn, crypts of, 156.
Lifting sections, 41.

Ligamenta subflava, 65.
Ligamentum nuchæ, 65.
Light transmitted, 9.
Limiting membrane, 49.
Linen fibers, 47.
Lines of Retzius, 137.
Littré, glands of, 196.
Liver, 163.
    cells of, 174.
    fat in, 174.
    Glisson's capsule, 163.
    glycogen in, 174.
    human, 171.
    of pig, 167.
    of rabbit, 174.
    practical demonstrations, 167, 170, 177.
    scheme of structure, 164.
Lobes of glands, 140.
Lobules of glands, 140.
Logwood, 27.
Loop of Henle, 181.
Lung, 123.
    air-sacs, 128.
    blood-vessels of, 128.
    connective tissue of, 128.
    fœtal, 134.
    infundibula of, 129.
    interlobular septa, 128.
    pigment within, 128.
    practical demonstration, 131.
    terminal bronchiole, 123, 132.
    vascular supply of, 128.
Luschka's tonsil, 149.
Lymph, 106.
    -corpuscles, 85, 106.
    -node, diagram of, 112.
    -node, practical demonstration of, 113.
    -nodes, 111.
    -spaces, 106, 107.
    -spaces of nerves, 219.
    -spaces around ganglion-cells, 219.
Lymphatic capillaries, 106, 110.
    vessels, 107.
    glands. See Lymph-nodes.
    system, 106.
    tissue. See Lymphoid tissue.
Lymphatics, 106, 107.
    practical demonstration, 108.
    perivascular, 107.
    valves of, 107, 109.
Lymphoid cells, 85, 113.
    tissue, 76, 111.
    tissue, diffuse, 113.
    tissue distribution, 113, 118, 121, 123, 149, 156, 159, 195.
Lymphocytes, 85.

Magnification of movements, 46.
Magnifying power of objectives, 7.
Male generative organs, 209.
Malpighian stratum of epidermis, 92.
Malpighian bodies of kidney, 180.
  bodies of spleen, 118.
Malpighi, pyramids of, 179.
Mammary glands, 208.
Marrow, bone-, 73.
Measurement of objects, 8.
Media of arteries, 103.
Mediastinum testis, 209.
Medulla of bone, 73.
  of kidney, 179.
  of lymph-node, 112.
  of hair, 95.
  of thymus body, 121.
Medullary cavity of bone, 74.
  rays of kidney, 179.
Medullated nerves, 216.
Meissner's plexus, 154, 160.
Membrana granulosa, 205. [branes.
  propria. See Basement mem-
Membrane, basement. See Basement membranes.
  glassy, 95.
  limiting, 49.
  of Nasmyth, 135.
  of Krause, 79.
Membranes of brain, 235.
Menstruation, uterus in, 197.
Mercuric chlorid method for nervous system, 34.
Mesenteric lymph-node, 113.
Mesentery, silvered, 61.
Mesoderm or Mesoblast, derivatives of, 54.
Methylene blue, 31, 84.
Methyl green, 32.
Methyl violet, 87.
Metric scale, 8.
Micro-millimeters, 8.
Micron, μ, 8.
Micrometers, 8.
Microscope, 1.
  adjustment of, 5.
  care of, 43.
  illumination, 4.
  magnifying power of, 7.
  parts of, 1.
  sketching from, 9.
Microtome, freezing, 19.
  Minot, 15.
  Schanze, 15.
  Stirling, 12.
  Thoma, 14.

Milk, 45, 208.
  colostrum-corpuscles, 208.
  fat-globules in, 45.
  secretion of, 208.
Mixed salivary glands, 148.
Mirror, 1, 4.
Mitosis, 51.
Molecular movement, 46.
Mounting fluids and methods, 35.
  of objects, 41.
Mounts, rings on, 36, 76.
Mouth, 149.
Movement, Brownian, 46.
  amœboid, 83.
  vital, 47.
Movements, magnification of, 46.
Mucoid tissue, 76.
Mucosa of bronchial tubes, 123.
  of stomach and intestine, 151.
  of mouth and pharynx, 149.
  of œsophagus, 150.
  ureter and bladder, 193.
  uterus, 199.
  vagina, 199.
Mucous glands, 127, 144, 148, 149.
  cells. See Goblet cells.
Müller's fluid, 22.
Multipolar nerve-cells, 220.
Muscular tissue, 76.
Muscle, blood-vessels of, 80.
  cardiac, 80.
  involuntary, 76. See Non-striated muscle.
  nerves of, 225.
  non-striated, 76.
  non-striated, distribution of, 76, 96, 102, 107, 112, 118, 123, 150, 151, 193, 194, 196, 197, 202.
  of hair-follicle, 96.
  smooth, 76. See Non-striated.
  striated, 78.
  voluntary, 78.
  of œsophagus, 150.
Myocardium, 80.

Naboth, ovules of, 199.
Nails, 98.
Nasal mucous membrane, epithelium, 55.
Nasmyth's membrane, 135.
Needles, 36.
Nerve-cells, 219.
  -cells of suprarenal body, 213.
  -cells, processes of, 220.
  -cells pyramidal, 238.
  -cells, types, 220

254　INDEX

Nerve-endings, 223.　[225.
　-endings in non-striated muscle,
　-endings in striated muscle, 225.
　-endings in skin, 94, 223, 224.
　-fibers, 216.
　-fibers, axis-cylinder of, 216.
　-fibers in osmic acid, 23.
　-fibers medullated, 216.
　-fibers non-medullated, 217.
Nerve, acoustic, 216.
　　connective tissue of, 218.
　　funiculi, 217.
　　olfactory, 217.
　　optic, 216.
　　plexuses, 225.
　　practical demonstration, 219.
　　spinal, 228.
　　-staining, Cox, 34.
　　-staining, Golgi, 33.
　　-staining, nigrosin, 32.
　　-staining, Van Gieson, 32.
　　-staining, Weigert's, 30.
　　-trunks, 217.
Nervi nervorum, 219.
Nervous system, 216.　[22, 23.
Nervous tissues, fixation or hardening,
　　tissues, development of, 54, 223.
　　tissues, supporting framework
　　　　of, 222.
Neurilemma, 217, 218.
Neuroglia, 222.
　-cells, 223.
Neurone, 221, 222.
Neutrophile granules, 31, 86.
Newt's tail for karyokinesis, 53.
Nigrosin, 32.
Nitrate of silver, 33, 61, 108.
Nitric acid, 24, 75.
Nodes, lymph-. See Lymphatic nodes.
　　of Ranvier, 216.
Non-medullated nerves, 217.
Non-striated muscle, 76. See Muscle,
　　non-striated.
Normal salt solution, 36.
Nose, epithelium of, 55.
Nuclei of cells, 49.
Nucleoli of cells, 50.
Nuclear stains, 27.
　　spindle, 52.
Nucleus, fibrils of, 49.
　　structure of, 49.

Objectives, 2, 4.
Ocular. See Eye-piece.
Odontoblasts, 135.

Œsophagus, 150.
Oil-immersion objective, 4.
　-globules, 46.
　of cedar wood, 4.
Oils, essential, 35.
Olfactory nerve, 217.
Optical axis, 5.
Optic nerve, 216.
Organisms in urine, 47.
Origanum oil, 35.
Orth's fluid, 22.
Osmic acid, 23.
Osmico-bichromate mixture, 23.
Ossification, 73.
Osteoblasts, 73.
Osteoclasts, 73.
Os uteri, 198.
Ovarial tubes, 206.
Ovary, 203.
　　corpus luteum of, 203.
　　germinal epithelium, 203.
　　Graafian follicles of, 203.　[206.
　　practical demonstrations of, 203,
　　tunica albuginea, 203.
Oviduct, 202.
Ovula Nabothi, 199.
Ovum, 205.
　　formation of, 206.
Oyster, ciliated cells from, 59.

Pacchionian bodies, 235.
Pacinian corpuscles, 223.
Pal-Weigert method, 30.
Pancreas, 145.
　　practical demonstration of, 146.
Paper, Japanese, 41.
Papillæ, circumvallate, 149.
　　filiform, 149.
　　foliatæ, 149.
　　fungiform, 149.
　　of skin, 93.
　　of tongue, 149.
Papillary eminences of kidney, 178.
Paraffin, cementing or soldering, 18.
　　imbedding, 13, 25.
Parenchyma of glands, 140.
Parotid gland, 143, 148.　[146.
　　gland, practical demonstration of,
Patch, Peyer's, 160.
Pavement epithelium, 56.
Pelvis of kidney, 193.
Penis, 212.
Peptic glands, 152.
Perforating fibers of Sharpey, 71.
Pericardium, 60, 102, 107.

INDEX

Perichondrium, 67.
Pericementum, 137.
Perimysium, 78.
Perineurium, 218.
Periosteal bone, 74.
Periosteum, 73.
Peripheral nerve-termini, 223.
Peritoneum, 60, 107, 151.
Perivascular lymphatics, 107. [107, 238.
    lymph-spaces of cerebrum,
Peyer's patches, 160.
Pharynx, 149.
    mucous membrane of, 149.
Pia mater, 235.
Picric acid, 23, 29, 31, 32.
    alcohol, 23.
Picro-carmine, 29.
Pigment-cells of hair, 93.
    -cells of skin, 93.
    -cells in suprarenal body, 215.
Pineal body, embryonic derivation, 54.
Pipettes for hæmocytometer, 86.
Pituitary body, embryonic derivation, 54.
Plates, blood-, 82.
    cartilage- in bronchi, 124.
Pleura, 60, 107, 128.
Plexuses, sympathetic, 225.
    of Auerbach, 154, 160.
    of Meissner, 154, 160.
Penciling of serous surfaces, 108.
Pole of the cell, 52.
Pollen, 48.
Polynuclear or polymorphonuclear leucocytes, 86.
Portal canal, 165.
    vein, 163.
Posterior commissure of spinal cord, 227.
    columns, 227.
    median fissure, spinal cord, 227.
Potassium bichromate, 22.
Potato starch, 48.
Power, magnifying, 7.
Practical demonstrations of—
    blood, 82-91.
    blood-vessels, 105.
    bone, 75.
    bronchial tube, 125.
    cartilage, 68.
    cerebellum, 240.
    cerebrum, 237.
    development of ovum, 206.
    elastic tissue, 65.
    endothelium, 60.
    epithelium, 55-60.
    Fallopian tube, 202.

Practical demonstrations of—
    hair, 99.
    intestine, 160-162.
    karyokinensis, 53.
    kidney, 186.
    liver, 167, 170, 177.
    lung, 131.
    lymphatics of central tendon of diaphragm, 108.
    mesenteric lymph-node, 113.
    mouth, œsophagus, pharynx, etc., 150.
    muscle, 77-80.
    nerves, 219.
    ovary, 203, 206.
    pancreas, 146.
    parotid gland, 146.
    skin, 98.
    spinal cord, 229.
    spleen, 118.
    stomach, 155.
    submaxillary gland, 146.
    suprarenal body, 213.
    teeth, 137.
    testicle, 211, 212.
    thymus body, 121.
    tongue and taste-buds, 150.
    ureter, 193.
    urinary bladder, 195.
    uterus, 197.
    vagina, 197.
    white fibrous tissue, 63.
Pregnancy, uterus in, 197.
Preservation of tissues, 20.
Prickle cells of skin, 92.
Prismatic color in air-bubbles, 46.
Prisms, enamel, 137.
Projection fibers, 239.
Prostate gland, 212.
Prostatic concretions, 212.
Protoplasm, 49.
    reticulation of, 50. [220
Protoplasmic processes of ganglion-cells,
Proximal convoluted tubule, 181.
Pseudopodia, 83.
Pulmonary alveoli, 128.
    artery, 128.
Pulp of teeth, 135.
    of spleen, 117.
Purkinje, cells of, 242.
Pus-corpuscles, 84.
Pyloric glands, 152.
Pyramidal tracts, 228.
Pyramid of Ferrein, 179.
    of Malpighi, 179.
Pyroxylin, 26.

Quick hardening with alcohol, 21.
    hardening with formaldehyde, 22.

Racemose glands, 143.
Ranvier's nodes, 216.
Rapid hardening with alcohol, 21.
    with formaldehyde, 22.
Rays, medullary, of kidney, 179.
Razor, form of, 10.
    stropping, 17.
Receptaculum chyli, 106.
Red blood-corpuscles, 81. See Blood.
    marrow, 73.
Reproductive organs, 197, 209.
Reptiles, blood of, 82.
Respiratory organs, 123.
Rete Malphighii of skin, 92.
    mucosum of skin, 92.
    testis, 210.
Reticular cartilage, 68.
    connective tissue, 76, 113. [113.
Reticulum of adenoid or lymphoid tissue,
Retiform tissue. See Reticular connective tissue.
Retzius, lines of, 137.
Ribbon sections. See Serial sections.
Ringing mounts, 36, 43, 76.
Root-sheath of hair, 95, 99.
Rugæ, 151.

Sacs, air-, 128.
Safranin, 23, 31.
Salamander tail for karyokinesis, 53.
Salivary gland, abdominal, 145.
    corpuscles, 47, 149.
    glands, 143, 148.
Salt solution, normal, 36.
Santorini, cartilage of, 123.
Sarcolemma, 78.
Sarcoplasm, 79.
Schanze microtome, 15.
Schwann, white substance of, 216.
Sebaceous glands, 98.
Sebum, 98.
Section cutting, 10.
    free-hand, 10.
    with microtomes, 12.
    in series, 15.
    -lifter, 41.
Sections, frozen, 19.
    serial, 15, 26.
    to place on slide, 41.
Seminiferous tubules, 211.
Sensory nerve-terminations, 223.
Serial sections, 15, 26.

Serous glands, 145, 148.
    membranes, 60, 107.
    membranes, lymphatics of, 108.
Sharpening knives, 16.
Sharpey's fibers, 71.
Sheath, dentinal, 135.
    of Schwann, 216.
Silk fibers, 47.
Silver, mesentery stained with, 61.
    nitrate, 33.
    staining, 33, 61, 108, 134.
    staining solution, 33.
Skeletal muscle, 78.
Sketching from microscope, 9.
Skin, arrector pili, 96.
    blood-vessels of, 94.
    corium, 92.
    chamois, 43.
    eleidin granules, 92.
    epidermis, 92.
    hair-follicles, 95.
    injected, 94.
    keratin of, 92.
    negro, 93.
    nerves of, 94.
    papillæ of, 93.
    pigment of, 93.
    practical demonstration, 98.
    sebaceous glands, 98.
    stratum corneum, 92.
    stratum granulosum, 92.
    stratum lucidum, 92.
    stratum Malpighii, 92.
    sudoriferous glands, 97.
    sweat-glands, 97.
    tactile corpuscles of, 94, 223.
Slides, 41.
Small intestine, 156.
Smooth muscle. See Muscle, non-striated.
Sole-plate, 225.
Solitary glands, 160.
Solution, Erlicki's, 24.
    Ehrlich-Biondi-Heidenhain, 32.
    Flemming's, 23.
    Müllers, 22.
    normal salt, 36.
    Orth's, 22.
    Stirling's, 25.
    Toison's, 87.
Space, subarachnoid, 235.
    subdural, 235.
Spaces, interglobular, 137.
    lymph-, 106.
    venous, 117.
Special connective tissues, 76.

INDEX

Specimens, permanent, 35.
Spermatogenic cells, 212.
Spermatozoa, 211.
Sphincter of renal papillæ, 193.
    of urinary bladder, 195.
Spider cells, 223.
Spinal cord, 226.
    anterior column, 227.
    anterior gray commissure, 227.
    anterior median fissure, 227.
    ascending antero-lateral tract, Burdach's column, 228. [228.
    central canal of, 227, 232.
    Clarke's column, 232.
    collateral fibers, 234.
    columns of, 227, 228.
    commissures of, 227, 228.
    crossed pyramidal tract, 228.
    descending antero-lateral tract, 228.
    direct cerebellar tract, 228.
    direct pyramidal tract, 228.
    funiculus cuneatus, 228.
    funiculus gracilis, 228.
    ganglion-cells, 232.
    Goll's column, 228.
    Gower's column, 228.
    gray commissures, 227.
    gray matter, 226.
    lateral columns, 227.
    nerve-fibers of, 231.
    nerve-roots of, 228.
    posterior column, 227.
    posterior median fissure, 227.
    practical demonstration, 229.
    staining of, 30, 32, 33, 229.
    substantia gelatinosa Rolandi, 232.
    substantia gelatinosa centralis, 232.
    Türck's column, 228.
    white commissure, 228.
    white matter of, 226.
Spiral tubule, 181.
Spleen, 117.
    Malpighian bodies, 118.
    practical demonstration, 118.
Spongy bone, 72.
Spot, germinal, 205.
Squamous epithelium, 55.
Staining agents, 27.
    aniline dyes, 31.
    borax-carmine, 29, 39.
    carmine, 29, 39.
    Cox-Golgi, 34.

Staining, double, 38.
    Ehrlich-Biondi-Heidenhain, 32, 85
    eosin, 28, 38.
    fresh tissue, 20.
    fuchsin, 31.
    general or ground, 27.
    gentian violet, 31.
    Golgi, 33.
    hæmatoxylin, 28, 36.
    hæmatoxylin and eosin, 38.
    in bulk, 29.
    methylene blue, 31, 84.
    nuclear, 27.
    nigrosin, 32.
    osmic acid in nerve tissue, 23.
    Pal, 30.
    picro-carmine, 29.
    selective, 27.
    silver, 33, 61, 108, 134.
    safranin, 23, 31.
    triacid, tricolor, triple, 32, 85.
    Van Gieson, 32.
    Weigert's nerve-, 30.
Starch, 48.
Stirling microtome, 12.
    dissociating fluid, 25.
Stomach, 151.
    practical demonstration, 155.
    lymphatics of, 156.
    mucous membrane of, 152.
    nerves of, 154.
    peptic glands, 153.
    pyloric glands, 154.
Stomata, 62, 109.
Stratified epithelium, 55.
Stratum corneum of skin, 92.
    granulosum, 92.
    lucidum, 92.
    Malpighii, 92.
Striated or striped muscle, 78. See Muscle, striated.
    cardiac muscle, 80.
Stripes of Baillarger, 237.
Stropping knives, 17.
Subarachnoidal space, 235.
Subdural space, 235.
Sublingual gland, 148.
Sublobular vein, 163.
Submaxillary glands, 144, 148.
    glands, practical demonstration, 146.
Substance, cement-. See Cement-substance.
    white, of Schwann, 216.
Substances, extraneous, 47.

Substantia, gelatinosa, 232.
Succus entericus, 156.
Sudoriferous glands, 97.
Sulphate of copper, 24.
Supernumerary spleens, 118.
Suprarenal body, 213.
    capsule, 213.
Sustentacular cells of testicle, 212.
Sweat-glands, 97.
Sympathetic system, 217.
System, cerebro-spinal, 226–243.
    circulatory, 102.
    digestive, 135–177.
    Haversian, 71.
    lymphatic, 106.
    nervous, 216.
    respiratory, 123.
    reproductive, 197, 209.
    sympathetic nervous, 217.

Tactile corpuscles, 94, 223.
Tangential fibers, 238.
Taste-buds, 149.
Teasing, 9.
Technology, 1.
Teeth, 135.
    cementum, 137.
    crusta petrosa, 135.
    decalcification of, 137.
    dentinal fibers, 135.
    dentinal tubules, 135.
    dentine, 135.
    enamel, 137.
    interglobular spaces, 137.
    membrane of Nasmyth, 135.
    odontoblasts, 135.
    pericementum, 137.
    practical demonstration of, 137.
    pulp, 135.
    stripes of Retzius, 137.
Teichmann's crystals, 90.
Tendon, 62.
Tesselated epithelium, 56.
Testicle, 209.
    coni vasculosi, 210.
    mediastinum of, 209.
    seminiferous tubules of, 210, 211.
    spermatozoa, 211.
    spermatogenesis, 212.
    tunica albuginea, 209.
    tunica vaginalis, 209.
    vasa efferentia, 210.
Thermal currents, 46.
Thoracic cavity, 60, 107, 128.
    duct, 106, 107.

Thymus body, 121.
Thyme oil, 35.
Thoma microtome, 14.
Thyro-glossal duct, 148.
Thyroid body, 148.
    body, colloid secretion of, 148.
Tissue, adenoid. See Lymphoid tissue.
    adipose, 65.
    areolar, 62.
    connective, 62.
    definition of, 53.
    elastic, 63.
    embryonic, 76.
    erectile, 212.
    fat-, 65.
    fibrous, 62.
    fixation of, 20.
    fresh, 20.
    hardening of, 20.
    lymphoid. See Lymphoid tissue.
    mucoid, 76.
    muscular, 76.
    nerve., 216.
    white fibrous, 62.
    yellow elastic, 63.
Tissues, classification of, 53.
    dehydration of, 35, 38.
    embryonic derivation of, 54.
    varieties of, 53.
Toison's fluid, 87.
Tongue, 149.
    muscle of, 79.
Tonsil of Luschka, 149.
Tonsils, 149.
Tooth. See Teeth.
Touch corpuscles, 94, 223.
Trabeculæ of lymph-nodes, 112.
    splenic, 117.
Trachea, 123.
    cartilages of, 68, 123.
Tracts of the spinal cord, 228.
Transitional epithelium, 55, 193, 195.
Tricolor or triacid stain of Ehrlich, etc., [32, 85.
True skin, 97.
Tube, bronchial, 123.
Tubular glands, 140.
Tubule, distal convoluted, 182.
    Fallopian, 202.
    gastric, 152.
    Henle's loop, 181.
    proximal convoluted, 181
    spiral, 181.
    straight, of kidney, 182.
    straight, of testicle, 210.
Tubules, seminiferous, 211.

INDEX

Tubules, uriniferous, 180, 182.
Tunica albuginea, 203, 209.
Türck, column of, 228.
Turpentine carbol-, 35.
Turn-table, 36.
Tunica vaginalis, 209.
       vasculosa, 210.
Typical cell, 49.

Unstriated or unstriped muscle, 76. See Muscle, non-striated.
Umbilical cord, 76.
Unipolar nerve-cells, 220.
Ureter, 193.
Urethra, 196.
      glands of, 196.
Urinary bladder, 194.
      organs, 178–196.
Urine, bacteria in, 47.
      course of, in kidney, 185.
Uriniferous tubules of kidney, 180.
Uterus, 197.

Vagina, 197.
Valves of heart, 102.
      of lymphatics, 107, 109.
      of veins, 104.
Van Gieson's stain, 32.
Vasa vasorum, 105.
Vascular system, 102, 106.
Vas deferens, 209.
Vasa efferentia, 210.
Vegetable fibers, 47.
      spores, 48.
Veins, 104.
      hepatic, 163.
      interlobular, 163.
      intralobular, 163.
      portal, 163.
      sublobular, 163.
      valves of, 104.

Venæ stellatæ of kidney, 184.
Venous spaces of spleen, 117.
Venulæ rectæ of kidney, 184.
Venules, 102.
Vermiform appendix, 162.
Vesicle, germinal, 205.
Vesicles, air-, 128.
Villi of intestine, 156.
Vital movements, 47.
Voluntary muscle, 78. See Muscle, striated.

Wall of cell, 49.
Wandering cells, 83, 104.
Weigert-Pal method, 30.
Welsbach gas-burner, 4.
Wharton's jelly, 76.
Wheat starch, 48.
White blood-corpuscles, 83, 104.
      commissure of spinal cord, 228.
      fibrous tissue, 62.
      matter of brain, 235.
      matter of spinal cord, 226.
      substance of Schwann, 216.
Wood fibers, 47.
Wool fibers, 47.
Wrisberg, cartilage of, 123.

Xylol, 35.
      balsam, 35.

Yeast, 48.
Yellow elastic tissue, 63.

Zinc cement, 36.
Zona fasciculata, 213.
      glomerulosa, 213.
      pellucida, 205.
      reticularis, 213.
      vasculosa, 203.

www.ingramcontent.com/pod-product-compliance
Lightning Source LLC
Chambersburg PA
CBHW031945230426
43672CB00010B/2059